The HACCP Food Safety Training Manual

Tara Paster

WILEY

John Wiley & Sons, Inc.

This book is printed on acid-free paper. ∞

Published by John Wiley & Sons, Inc., Hoboken, New Jersey
Published simultaneously in Canada

For general information on our other products and services or for technical support, please contact our Customer Care Department within the United States at 800-762-2974, outside the United States at (317) 572-3993 or fax (317) 572-4002.

Wiley also publishes its books in a variety of electronic formats. Some content that appears in print may not be available in electronic books. For more information about Wiley products, visit our web site at www.wiley.com.

Library of Congress Cataloging-in-Publication Data:

Paster, Tara, 1968-
   The HACCP food safety training manual / Tara Paster.
       p.   cm.
   Includes index.
   ISBN-13: 978-0-471-78448-7 (pbk.)
   ISBN-10: 0-471-78448-6 (pbk.)
   1. Food industry and trade—Safety measures.   2. Food handling—Handbooks, manuals, etc.   3. Foodservice—Handbooks, manuals, etc.   I. Title.
   TX537.P292 2006
   664.00289—dc22

                                       2005035004

Printed in the United States of America

10 9 8 7 6 5 4 3 2 1

Contents

HACCP STAR POINT 1
PREREQUISITE PROGRAMS   1

ACKNOWLEDGMENTS

Just as you create and execute an effective HACCP Plan with your staff, it has taken the help of a TEAM to complete *The HACCP Food Safety Training Manual*! I would like to recognize the Superstars on my team who helped me to complete this exciting project.

Special Acknowledgment to:

Fay Algeo: Thank you for your expertise in training, communication, organization, feedback, and the flow of the book. You are very gifted with the ability to make challenging material fun and educational. As a Professional Trainer for the hospitality industry, your field experience with Paster Training, Inc. came across in the recommendations that you made from start to finish. All the hours and the tremendous job you did in our second testing of the book to food handlers made a great impact on the entire *HACCP Food Safety Training Manual*.

Carol Gilbert: Thank you for the HACCP expertise you brought to this program. The contributions you made from your perspective as Food Service Director for Hempfield School District has really made a difference. You provided a different perspective, enabling us to edit this book to fit the needs of schools throughout the world.

Tony Paster: Words cannot describe the gratitude I have for my husband with regard to this HACCP project. The technical support you gave me in the creation of forms and documents and the marathon of editing adventures we traveled is overwhelming. I appreciate your support, dedication, and commitment to me and this HACCP project. I love you and thank you for everything!

Misty Doane: Thank you for the time you dedicated to this project. Your writing talent, input, and research helped facilitate the production of this book.

Extra Special Thanks Goes to:

JoAnna Turtletaub: Thank you for giving me this opportunity to impact the school foodservice and the hospitality industry with HACCP training material for employees, managers, and instructors throughout the world.

Nigar Hale, Julie Kerr, and Cindy Rhoads: This is the dynamic trio at Wiley who led me through this electrifying process. I have a huge appreciation for their commitment and genuine interest in this project. Thank you for consulting and monitoring the new and stimulating manager's HACCP book.

The Wiley Production Team: WOW! The Wiley Production Team, including all the editors, designers, compositor, and artists, thank you all for a job well done.

The entire Manager HACCP Writing TEAM (Fay, Carol, Misty, Tony, JoAnna, Nigar, Julie, Cindy, and the Wiley Production Team) is an incredible group of professionals who deserve special acknowledgment and recognition for their outstanding efforts!

HACCP Introduction: Star Points to Food Safety

No matter where you are in the world, on a clear night you can look up in the sky and see millions of sparkling stars in our solar system. Each of these sparkling stars is unique and different, just like the millions of foodservice operations of the world. Each foodservice operation is unique, whether a school or another institution, independently owned or part of a franchise. That is why every operation serving or selling food needs to have a food safety system in place that is uniquely designed to guarantee that the food being served is safe to eat. This specific food safety system is called **HACCP** (pronounced has-sip), or **Hazard Analysis** and **Critical Control Point**. HACCP is a system composed of seven principles that are meant to be applied to a written food safety program focusing on the food in your operation.

It includes **prerequisite programs**, which are basic operational and foundational requirements needed for an effective HACCP plan. Prerequisite programs covered in this book include the following:

Introduction

  ★ Training—Employee training and manager accreditation
  ★ Sanitation
  ★ Active managerial control

Star Point 1: Prerequisite Programs

  ★ Product instructions (recipe and process)
  ★ Equipment
  ★ Facility design
  ★ Standard operating procedures
  ★ Supplier selection and control
  ★ Product specifications (purchasing)

- ★ Personal hygiene/employee health
- ★ Allergen management
- ★ Food safety
- ★ Chemical and pest control

Star Point 2: Food Defense

- ★ Food defense
- ★ Food recall procedures
- ★ Crisis management

HACCP proves that what you do or *don't* do makes a big difference in serving safe food. The **goal of HACCP** is to be proactive by stopping, controlling, and preventing food safety problems using prerequisite programs and the seven HACCP principles. The goal of this book is for you to become a HACCP Superstar and earn your HACCP certification!

The HACCP system is very important because it saves lives! The CDC (Centers for Disease Control and Prevention) estimates that every year, **76 million** people get sick from eating unsafe food. Out of those millions of people, **325,000 people are hospitalized** and **5,000 people die** from eating unsafe food. The HACCP system requires the management team to provide solid commitment, strong leadership, and adequate resources to the HACCP program to prevent these tragedies. Every team member in the foodservice industry must be **responsible** to ensure that the food he or she prepares and serves to customers is not hazardous to their health. The CDC has identified the top five reasons why food becomes unsafe, known as **foodborne illness risk factors**. The foodborne illness risk factors identified help management focus on specific *proactive* food safety goals for each foodservice establishment, ultimately achieving **active managerial control**. The foodborne illness risk factors are as follows:

- ★ Poor personal hygiene
- ★ Not cooking food to the minimum cooking (internal) temperature
- ★ Not holding food properly
- ★ Cross-contamination of food, equipment, and utensils
- ★ Purchasing food from unsafe suppliers

The star points to food safety covered in this book specifically address these top five causes of unsafe food and how to prevent them. Prevention is achieved through active managerial control in the form of food safety management systems such as prerequisite programs with an emphasis on developing and implementing **standard operating procedures** (SOPs) and applying the seven HACCP principles. The expectation is for management to take the sample SOPs, charts, and record-keeping forms in this book and customize them for their foodservice operation and work to achieve active managerial control. The 2005 FDA Model Food Code defines **active managerial control** as the "purposeful incorporation of specific actions or procedures by industry management into the operation of their business to attain control over foodborne illness risk factors. It embodies a *preventive* rather than *reactive* approach to food safety through a continuous system of monitoring and verification." Using the SOPs as a starting point or checklist enables you to compare the recommendations in this book with your existing operation, which will result in a **needs assessment**. The needs assessment will provide you with some actions that need to be taken in your foodservice operation to ensure the safe preparation and service of food. The key words here are achieve, active, and action. As you can see, *it takes energy to improve your business!*

This book is intended to motivate you to **improve business**. Your business will increase if the quality of your food and the skills of employees can enhance the operation's reputation. The HACCP plan, including establishing prerequisite programs and achieving active managerial control, will do this for you. When HACCP is properly used by schools and businesses, worldwide food safety improves and fewer people die. Almost every one of the 5,000 deaths that occur every year from eating unsafe food could have been prevented. We know how to make food safe through the use of HACCP and by achieving active managerial control!

Besides the moral obligation, revenue will increase if you use HACCP. You will make more money because your employees are better trained and more efficient. They are more aware and focused on the food. Focusing on the safety of food products naturally creates a more consistent food product that leads to an added bonus of exceptional food quality. With a HACCP system, every ingredient is important and every process is documented, which results in an increase in product quality, tighter controls, improved food cost, and a reduction in product loss. Better food cost gives you more profit because you are managing and controlling your business ingredient by ingredient.

Schools, retail businesses, grocery stores, convenience stores, mobile units, and institutional, independent, and franchise foodservice operators face many challenges every day in implementing prerequisite programs and the seven HACCP principles. This book would be remiss if these challenges were not recognized. As an operator, view these challenges as speed bumps. It may be necessary to slow down and move carefully over the bump. The same is true for dealing with the challenges that face your foodservice operation; at times the process can be slow. To manage the situation, you need a carefully thought-out plan. The key is to keep moving forward and work the course. Some of the challenges and speed bumps that face foodservice operators include the following:

- ★ Limited financial resources—capital needed to properly operate
- ★ Large number of menu items and products
- ★ Frequently changing menus and procedures
- ★ Inadequate organizational structure and support
- ★ Employee turnover
- ★ Multicultural workforce
- ★ Varied educational levels
- ★ Communication (language barriers)
- ★ Implementation of regulatory requirements/laws

However big these challenges may be at times, none of these should serve as excuses for poor execution in the day-to-day operation of your facility. Nor should these be reasons why the prerequisite programs and seven HACCP principles are not achieved by properly trained operators and their team members.

Finally, HACCP forces you to be involved in all the day-to-day activities of your facility and to identify and document areas of needed improvement. It requires you to participate, take action, and achieve the goals of HACCP. When food safety, quality, and consistency are improved, your customer traffic should increase, which then increases your sales. Increased sales give you more opportunity to increase your profitability.

Prerequisite programs and food defense standards of operation are the building blocks for creating an effective HACCP plan—if any member of the foodservice operation **does not** follow these standard operating procedures, even the most well thought out HACCP plan will fail. To ensure the development of an effective HACCP plan for your establishment, you must review the basics of food safety and food defense standard operating procedures. We will cover these in the first two chapters. Once the basics of food safety and food defense standard operating procedures are reviewed, the book then focuses on how a HACCP plan is created and **how to use an effective HACCP plan** for your foodservice establishment.

★ THE HACCP PHILOSOPHY

HACCP is internationally accepted. It is critical to note that it is not a process conducted by an individual, it involves the entire **team,** which is why you are a part of this training session. We are counting on you to do your part in preventing foodborne illness in your foodservice operation and in your part of the world. Every foodservice facility must have leadership; if you are responsible for any part of the operation, then you need to **demonstrate effective leadership skills.** As a leader, ask yourself these questions:

- ★ Can I be a role model for food safety and HACCP?
- ★ Can I provide support to the HACCP team?

If you are the top leader/manager in your organization, here are the additional questions that you need to ask yourself:

- ★ Can I provide strong leadership for my HACCP team?
- ★ Have I assembled the best-qualified HACCP team possible?
- ★ Do I encourage and expect my HACCP team to implement the best food safety and HACCP plan possible?
- ★ Can I provide the necessary resources for my HACCP team?

As a leader, if you are not a proper role model and do not provide solid direction to your team, the time and money that you have invested is a financial loss. Worse yet, death could occur, resulting in the destruction of your business as well as your brand. **Leadership** is all about making intelligent and informed decisions. Based on your decisions, your HACCP plan will fail if foodservice leaders do not support good food safety practices and are not proactive and resourceful in creating and implementing your HACCP plan.

The **HACCP philosophy** simply states that biological, chemical, or physical hazards, at certain points in the flow of food, can be

- ★ Prevented
- ★ Removed
- ★ Reduced to safe levels.

Today, foods are transported around the world more than ever before. As a result, more people are handling food products. The more food products are touched by people or machines, the greater the possibility for contamination or, even worse, the spread of a foodborne illness.

The eating habits of people around the world have changed as well. People eat more ready-to-eat foods and enjoy more ethnic dishes and food varieties than ever before. This is yet another reason why we need HACCP!

★ OUR MISSION

To minimize consumer risk of contracting a foodborne illness, to prevent having an allergic reaction to food, and to avoid suffering an injury from foods consumed in any foodservice operation around the world is the mission of a HACCP program.

■ THE HACCP STAR

The goal of this HACCP training program is to make you a **HACCP Superstar!** To be a HACCP Superstar, you must shine on all five points of the HACCP Star. Here is the HACCP Star and the five major points that define a successful HACCP system.

> ★ **Establish Prerequisite Programs**
> ★ **Apply Food Defense**
> ★ **Evaluate Hazards and Critical Control Points**
> ★ **Manage Critical Limits, Monitoring, and Corrective Actions**
> ★ **Confirm by Record Keeping, Documentation**

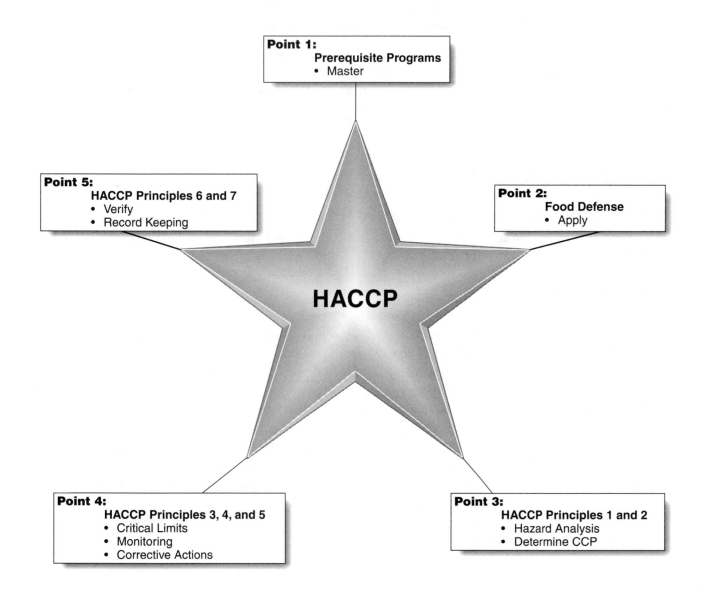

Point 1:
  Prerequisite Programs
  • Master

Point 5:
  HACCP Principles 6 and 7
  • Verify
  • Record Keeping

Point 2:
  Food Defense
  • Apply

HACCP

Point 4:
  HACCP Principles 3, 4, and 5
  • Critical Limits
  • Monitoring
  • Corrective Actions

Point 3:
  HACCP Principles 1 and 2
  • Hazard Analysis
  • Determine CCP

Your HACCP certification expires in 4 years, so it is critical to keep your certification current. Always reach for the stars! Once you read this entire book you will be able to

★ **Explain the Codex Alimentarius Commission.**

★ **Identify the causes of most foodborne illnesses.**

★ **Apply FDA and USDA initiatives to your foodservice operation.**

★ **Explain the seven HACCP principles.**

★ **Identify the key points of HACCP.**

★ **Follow standard operating procedures for food safety.**

★ **Apply standard operating procedures for food defense in your operation.**

★ **Explain how to conduct a hazard analysis.**

★ **Identify three classifications of recipes.**

★ **Determine critical control points.**

★ **Apply correct critical limits.**

★ **Complete monitoring forms.**

★ **Determine effective corrective actions.**

★ **Explain the verification process.**

★ **Apply documentation and record keeping to your operation.**

To start, take a **HACCP Pretest** to measure your current food safety, food defense, and HACCP knowledge. This pretest allows your trainer to measure your success as you work toward your HACCP Superstar Certificate. Let's get started.

★ HACCP PRETEST

1. The Codex Alimentarius Commission was created by the

   a. World Health Organization and the Food and Agriculture Organization of the United Nations.
   b. National Advisory Committee on Microbiological Criteria for Foods.
   c. U.S. Food and Drug Administration and the U.S. Department of Agriculture.
   d. Department of Homeland Security.

2. Conducting a hazard analysis means

   a. "What is the likelihood of a hazard to occur?" *and* "What are the standard operating procedures?"
   b. "What is the likelihood of a hazard to occur?" *and* "What is the risk if the hazard does occur?"
   c. "What is the risk if the hazard does occur?" *and* "What is the rational means of ensuring critical control points and verification do occur?"
   d. "What is active managerial control?" *and* "Analyze the hazards of operating procedures."

3. A critical control point (CCP) is

   a. An essential step in the product-handling process where controls can be applied and a food safety hazard can be prevented, eliminated, or reduced to acceptable levels
   b. Preventing problems in the corrective action, verification, and record-keeping processes
   c. Monitoring hand washing when changing tasks
   d. The common goal of operators and regulators of retail and foodservice establishments to produce safe, quality food for consumers

4. Record keeping includes

   a. The severity of a biological, chemical, and physical hazard
   b. Employee training classes
   c. Checking for critical limits
   d. FDA Forms 1-A, 1-B, and 1-C

5. Prerequisite programs are

   a. Several conditions documenting the hazards before the CCP
   b. Cooking food to its proper temperature
   c. Basic operational and foundational requirements that are needed for an effective foodservice HACCP plan
   d. The inspection reports of the local regulatory agency

6. Critical limits can be

   a. Cleaning food-contact surfaces
   b. Cooking foods to a specific temperature for a specific amount of time
   c. The maximum amount of time an employee can handle dangerous chemicals
   d. The amount of mold that is safe to eat

7. If cooling is the CCP, then the critical limit is

   a. 135°F to 41°F (57.2°C to 5°C) in more than 4 hours
   b. 135°F to 70°F (57.2°C to 21.1°C) within 4 hours and 70°F to 41°F (21.1°C to 5°C), with an additional 2 hours
   c. 135°F to 70°F (57.2°C to 21.1°C) within 4 hours and 70°F to 41°F (21.1°C to 5°C), with an additional 4 hours
   d. 135°F to 70°F (57.2°C to 21.1°C) within 2 hours and 70°F to 41°F (21.1°C to 5°C), with an additional 4 hours

8. What are the three classifications a menu is divided into during a hazard analysis?

   a. Ready-to-eat/convenience, full-service, and USDA commodity food
   b. No-cook/simple, same-day, and complex
   c. Ready-to-eat, USDA commodity food, and complex
   d. Appetizer, entrée, and dessert

9. Monitoring procedures involve

   a. Ensuring that we are correctly meeting critical limits for the CCPs
   b. Training employees in preventing cross-contamination
   c. Analyzing how disposed food affects the profit/loss statement
   d. Installing security cameras outside the establishment

10. What is food defense?

   a. Cooking food to the proper temperature to defend against pathogens
   b. Not allowing customers to enter the foodservice operation
   c. A new federal office that reports to the Department of Homeland Security
   d. The idea of preventing the deliberate contamination of food

11. An example of a corrective action is

   a. Issuing a written warning
   b. Showing a coworker how to work more efficiently while preparing food
   c. Rejecting a product that does not meet purchasing or receiving specifications
   d. Using FDA Form 1-C to allow a formerly ill employee to return to work

12. Which is not a form of verification for a HACCP plan?

   a. Checking equipment temperatures
   b. Checking critical control point records
   c. Making sure employees wear clean uniforms
   d. Point system for cleaning defects

13. What is the temperature danger zone?

   a. 45°F to 140°F (7.2°C to 60°C)
   b. 35°F to 140°F (1.7°C to 60°C)
   c. 41°F to 135°F (5°C to 57.2°C)
   d. 41°F to 165°F (5°C to 73.9°C)

14. What are the characteristics of potentially hazardous foods (PHFs)?

   a. Dry, low acidity, vegetable based
   b. Moist, neutral acidity, protein
   c. Moist, sugary, low fat
   d. Moist, vegetable based, high fat

15. What is food security?

   a. A 2-year supply of food for a country
   b. Designating an employee to watch the buffet
   c. A newly appointed government office
   d. Keeping food properly wrapped in storage

How many points did you earn? _____

If you scored 14–15 points—Congratulations! You are very knowledgeable already about HACCP!

If you scored 9–13 points—Good job! You have a basic understanding of HACCP and all of its components.

If you scored 5–8 points—There is no time like the present to learn about HACCP! This book will give you a great opportunity to fine-tune your HACCP skills.

If you scored 0–4 points—Everyone needs to start somewhere! It is important to track your progress as you complete each point of the star to earn your HACCP Superstar certification!

■ TRAINING

The emphasis to your team members and to your organization should be training and establishing the understanding that what your team does or does not do is significant to public health. Your training in food safety could actually save lives and help raise the quality of food served at your establishment. An effective HACCP plan includes training as a prerequisite program and as an essential component to your HACCP plan. As a leader and coach, you need to perform the following actions in your foodservice operation:

   ★ Check your local regulatory agency for certification requirements for person-in-charge/manager accreditation.

   ★ Develop an organizational chart and job descriptions showing assigned responsibilities for prerequisite programs and HACCP.

   ★ Set food safety and food defense goals that are challenging, measurable, and achievable.

   ★ Establish accountability for meeting food safety and food defense responsibilities.

★ Reinforce and recognize success with incentives and awards.

★ Demonstrate management's commitment through correct food safety and food defense behaviors—be a positive role model and always set the example.

★ Implement ongoing self-inspection and third-party inspection programs.

★ Encourage all team members to alert the person in charge to any food safety and food defense concerns immediately.

The basic training components of an effective HACCP plan should

★ Explain the training system and the process used to achieve effective and satisfactory job performance.

★ Assess training needs of all team members and every level in your organization, such as chefs, servers, or maintenance crew.

★ Provide training, knowledge, and technical skills instruction prior to all new job assignments.

★ Utilize outside/third-party training companies as needed to reinforce management's commitment to food safety and food defense behaviors (i.e., nationally recognized programs for manager and employee levels).

Post-training components should include

★ Updating training materials/procedures at least once a year

★ Conducting ongoing 20-minute sessions using demonstrations and hands-on activities to reinforce acquired skills

★ Encouraging all team members to give feedback as to how to improve training

★ Maintaining training records or charts to include the topic, materials, date, length of time, who attended, and trainer name

Most importantly, you can make a difference by following prerequisite programs such as good food safety practices, known as **standard operating procedures** (SOPs), and by making sound decisions that will help keep your customers safe. Approved HACCP plans require that each employee follow prerequisite programs and SOPs at each step in the flow of food. These are the standards you must know and practice when purchasing, receiving, storing, preparing, cooking, holding, cooling, reheating, and serving food. Job descriptions should make it clear that all employees must follow standard operating procedures. Here is a sample job description of a kitchen manager provided by The Food Experience.™

JOB DESCRIPTION

Date:	February 2005	Status:	Full-Time
Job Title:	Kitchen Manager	Reports to:	Owner/General Manager
Bonus:	Eligible—see bonus program for details	Location:	Collegeville, PA

Job Purpose

The Kitchen Manager (KM) functions as the person that will be leading the day-to-day operations of The Food Experience™. The primary role is development, implementation, and communication of company product and service in accordance with the company mission statement, corporate philosophy, values, and food safety standards. Focus is on meeting and exceeding consumer expectations while ensuring consumer retention via superior service, menu/dinner instructions, and operations management.

Job Responsibilities

This is a list of the major responsibilities and duties required of the Kitchen Manager position.

★ Manage consumer relations, vendor relations, shipments/deliveries, and all ordering.

★ Collaborate with company's business owners to ensure successful operations and customer satisfaction.

★ Communicate with consumers and build *The Food Experience*™ Brand name and approach when necessary.

★ Attend Paster Training Food Safety and Sanitation Program to become a certified food manager. Attend and successfully complete the class, remaining certified in ServSafe® throughout your employ with The Food Experience™.

★ Understand, communicate, and implement prerequisite programs and standard operating procedures, and follow the HACCP plan.

★ Determine product components including: menu items, ingredients, execution, and hard costs.

★ Manage daily operations of the kitchen facility, including:
  ★ Employees (hiring, management, schedules, communication and work)
  ★ Receiving
  ★ Inventory management
  ★ Prepping food

- ★ Prepping meal ingredients
- ★ Session prep and setup
- ★ Meal station assembly
- ★ Meal station breakdown
- ★ Customer interactions (sessions, calls, inquiries, tours, special requests)
- ★ Restocking inventory
- ★ Washing and washroom duties
- ★ Implementing, enforcing, and communicating food safety guidelines, and standards operating procedures
- ★ Train and support franchisees when applicable.
- ★ Prep ingredients, workstations, and retail outlet for sessions/meal development.
- ★ Perform project management duties including maintaining good vendor relationships, food ordering/tracing, and internal and external reporting.
- ★ Maintain food areas and operations in accordance with Health Department Regulations and FDA food safety guidelines.
- ★ Other related duties as assigned.

Required Qualifications

- ★ Must have 2 to 4 years of management experience
- ★ Must have 4 or more years foodservice experience
- ★ Must be a creative self-starter
- ★ Must have strong organization, negotiation, and problem-solving skills
- ★ Must have excellent communication skills, both written and verbal
- ★ Must be able to present effectively to small and medium-sized groups
- ★ Must be able to effectively handle multiple tasks and projects simultaneously
- ★ Must be a team player with great people skills
- ★ Must have proven work experience and references
- ★ Must be reliable, honest, trustworthy, hardworking
- ★ Must have experience with Back-of-House and Front-of-House operations
- ★ Must have basic computer skills (ability to utilize Internet, company intranet, and various software)

Physical Demands

In an average workday, associate would perform the following:

	Never (0% of shift)	Occasionally (up to 33% of time)	Frequently (33-66%)	Continuously (66-100% of shift)
Sit		X		
Stand			X	X
Walk			X	
Bend and/or stoop			X	X
Crawl and/or climb		X		
Kneel		X		
Push			X	
Pull		X		
Work on unprotected heights	X			
Operate mechanical machinery		X		
Be exposed to marked changes in temperature and humidity		X	X	
Be exposed to harmful fumes and/or other pollutants	X			
Use feet and/or legs for repetitive motion Left Right			X X	
Use hands and/or legs for repetitive motion Simple grasping Firm grasping (pushing pulling arm controls) Fine manipulation				X X X
Lifting and/or carrying Up to 10 pounds (sedentary level) 11–20 lbs. (light work level) 20–50 lbs. (medium work level) 52-74 lbs. (heavy work) 75-100 lbs. (heavy work)		X X	X	X X

A job description should not be vague or obtuse. The accountability of accepting their **role** in the organization, their **responsibilities**, and **physical demands** essential to the success of the foodservice operation should be clear to any qualified candidate. The reason for showing the preceding job description in such detail is to outline the **expectations** for team members to meet and hopefully exceed food safety and sanitation requirements.

Now let's take a look at the difference between food safety and sanitation.

⭐ FOOD SAFETY VS. SANITATION

With the constant battle of science and technology against the threat of emerging pathogens, the management of food safety is constantly changing. Every day foodservice managers are at war with an enemy that often cannot be seen, tasted, or smelled. **Food safety** involves keeping food safe to eat at every stage of handling as it passes through the flow of food from farm to table (purchasing, receiving, storing, preparing, cooking, holding, cooling, reheating, and serving). **Sanitation** is making sure anything that comes in contact with food at any stage of handling does not contaminate the food. Sanitation also involves pest control, equipment maintenance, and proper cleaning and sanitizing techniques. Sanitation is a prerequisite to food safety. However, you cannot have one without the other. Simply keeping things clean does not necessarily lead to food safety.

Traditional sanitation systems rely on observing dirt, then removing it. Food safety goes beyond what you can see. You should make a step forward to have a proactive food safety management system, known as "achieving active managerial control."

⭐ STAR KNOWLEDGE: ACTIVE MANAGERIAL CONTROL

Assess whether your food safety management system is up-to-date. Answer yes or no to the questions regarding implementing the following items in your foodservice operation.

Analyze Your Food Safety Management System	Yes	No
1. Do you have an enforced policy and procedure to ensure proper hand washing?		
2. Do you have an enforced policy to determine when employees are sick or have flulike symptoms? Do you use FDA forms 1-A, 1-B, and 1-C?		
3. Do you pay more attention to food temperatures than to the cleanliness of your facility?		
4. Do you have an ample supply of thermometers accessible to all employees throughout your operation?		
5. Do you calibrate thermometers every shift?		
6. Do you monitor the process, take corrective actions, verify, and maintain record keeping proving the food is cooled properly?		
7. Do you monitor the process, take corrective actions, verify, and maintain record keeping proving the food is reheated properly?		
8. Do you inspect your suppliers?		
9. Do you know if your suppliers have prerequisite programs with food safety and food defense standard operating procedures, a HACCP plan, and documentation that proves their food is safe?		

(continues)

Analyze Your Food Safety Management System	Yes	No
10. Do you have chemicals and food delivered on separate trucks or pallets? Do you know without a doubt that a chemical contamination has not occurred?		
11. Do you know what the correct minimum cooking temperatures are according to the 2005 FDA Model Food Code? Do you cook food to the correct minimum temperatures? Do you monitor the cooking process, take corrective actions, verify, and maintain record keeping proving the food is cooked properly?		
12. Do you hold food correctly? Do you monitor food cold- and hot-holding on a continuous basis, take corrective actions, verify, and maintain record keeping proving the food is held properly?		
13. Do you know at which critical control points (steps) in your food preparation system you are at highest risk for cross-contamination?		
14. Do you have procedures in place to prevent cross-contamination of equipment and utensils? Do you clean and sanitize all food-contact surfaces a minimum of every 4 hours? Do you have a system to test your sanitizer solution, verifying the concentration each time you fill a three-compartment sink, fill a bucket, or make a spray solution? Do you have a system to verify that the sanitizer is being properly used?		
15. Do you create an environment that prevents the deliberate contamination of food? Do you train your employees on food defense standard operating procedures for fellow coworkers, customers, vendor, and facility awareness?		
Tally the total number of "Yes" or "No" responses		

If you answered no to *any* of the preceding assessment questions, it is highly recommended that you update your current food safety management system and apply active managerial control. If you answered yes to *all* of the preceding questions, this is validation that your foodservice operation is focused on achieving active managerial control.

■ ACHIEVING ACTIVE MANAGERIAL CONTROL

The new way of thinking is to achieve active managerial control of foodborne illness risk factors. Active managerial control means the purposeful incorporation of specific actions or procedures by industry management into the operation of their business to obtain control over foodborne illness risk factors. It embodies a preventative rather than reactive approach to food safety through a continuous system of monitoring and verification. Five factors are known to cause **80 percent** of foodborne illness outbreaks:

★ Poor personal hygiene
★ Not cooking food to the minimum internal cooking temperature
★ Not holding food properly
★ Cross-contamination of equipment and utensils
★ Purchasing food from unsafe suppliers

HACCP (Hazard Analysis and Critical Control Points) is a food safety system that focuses on **potentially hazardous foods** and **time/temperature control for safe food**—and how they are handled in the foodservice environment. Self-inspection is the critical ingredient in HACCP. The basic HACCP concepts do support retail with TQM (total quality management) strategies. This system uses a flow chart to identify steps that are likely to cause failure in a process and to develop procedures to lower risks.

In order for a successful HACCP program to be implemented, management must be committed to HACCP. A commitment by management indicates an awareness of the benefits and costs of HACCP, which includes education and training of employees. Benefits, in addition to food safety, are a better use of resources and timely response to problems.

The focus of a HACCP inspection is on how food is handled, not on aesthetics. The result is safer food handling and, consequently, safer food.

Prerequisite Programs

Point 1:
Prerequisite Programs
- Master

HACCP

In the Introduction, the importance of prerequisite programs like training, sanitation, and active managerial control were discussed to give you insight to this first section of the book. In Star Point 1, we continue to discuss the basics of prerequisite programs (product instructions, equipment, and facility design) and food safety standard operating procedures (SOPs) using the International Food Safety Icons. We must be aware of the causes for unsafe food. We must also have rules and procedures in place to prevent the food from becoming unsafe. The prerequisite programs and established standard operating procedures can then be incorporated as part of the foundation for your HACCP plan.

Star Point Actions: You will learn to

★ Develop prerequisite programs.

★ Use prerequisite programs (product instruction, equipment, facility).

★ Recognize and understand the importance of SOPs.

★ Identify the causes for foodborne illness.

★ Describe how HACCP controls foodborne illness outbreaks.

★ Explain the transition between potentially hazardous food (PHF) and time/temperature control for safety of food (TCS).

★ Assist customers who have food allergies.

★ Identify the International Food Safety Icons.

★ Apply time and temperature controls to ensure food safety.

★ Prevent contamination of food.

★ Explain the personal responsibilities of each HACCP team member with regard to food safety.

★ Explain the difference between cleaning and sanitizing.

★ DEVELOPING PREREQUISITE PROGRAMS

The purpose for having food safety prerequisite programs in place is to control bacterial growth, protect products, and maintain equipment. The other benefits that you receive are customer satisfaction that result in increased customer counts and ultimately increased sales; employee satisfaction because you have designed a safe and easy working environment, resulting in increased productivity and money savings on labor; and energy savings because your foodservice facility is designed for efficiency and profit making.

Based on your operation these additional prerequisite programs will need to be reviewed before you start to implement an effective HACCP plan:

★ **Chemicals and Pest Control.** Do you have a secured locked location for chemicals? Are employees trained in the use of chemicals? Do you have an integrated pest management program with pesticides applied by a licensed pest control operator?

★ **Personal Hygiene/Employee Health.** Do you have a written dress code? Do employees follow the dress code? Do employees know the procedures for working when ill?

★ **Supplier Selection and Control.** Do suppliers have an effective HACCP plan? Do suppliers have effective food safety programs in place? Do suppliers apply food defense to their operation?

★ **Product Specifications.** Are specifications written for all ingredients, products, and supplies?

★ **Training.** Are employees receiving training in prerequisite programs, especially those related to their job duties such as personal hygiene, cleaning, sanitizing, and food safety?

★ **Food Safety.** Are procedures written and established for proper monitoring of food temperatures, cooling food, and reheating foods?

★ **Allergen Management.** Are employees aware of the primary food allergens? Do they know how to respond to customers' concerns regarding allergy questions?

Let's explore the prerequisite purpose in greater detail. The logical progression discussed in the Introduction showed that training your team is crucial, explained the difference between food safety and sanitation, and clarified why active managerial control is essential. If food can be time/temperature abused, then it is essential to control bacterial growth through such standard operating procedures as proper cooking and holding. If food is contaminated with biological (bacteria, viruses, and parasites), chemical (cleaning chemicals, pest control supplies, etc.), or physical (dirt, hair, glass, etc.) hazards, then prerequisite programs are used to protect the products from any of these contaminants. For example, following cleaning and sanitizing standard operating procedures is one step in preventing contamination.

Equipment is fundamental to the success of your foodservice operation prerequisite programs. To determine the necessary equipment needed, you must first decide on your menu, then the recipes and processes you will use. After you decide on the food products and required equipment, you can decide what type of facility you need to safely produce foods in the operation. This is why regulatory officials require a plan review before construction begins in a new foodservice facility. They are verifying that you have completely thought out your concept before the first nail is hammered. The next phase is product instructions (recipes), equipment, and facility design.

■ PRODUCT INSTRUCTIONS

Now that you clearly understand your concept, developed your menu, created recipes, and decided on the process and product instructions, it is time to apply it to the products you will offer. An example of this is in the recipe and instructions for mixed-fruit crisp.

MIXED-FRUIT CRISP	
1 15-ounce can (443.6 ml)	mixed fruit
1/2 cup (118.29 ml)	quick rolled oats
1/2 cup (118.29 ml)	brown sugar
1/2 cup (118.29 ml)	all-purpose flour
1/4 teaspoon (1.24 ml)	baking powder
1/2 teaspoon (2.45 ml)	ground cinnamon
1/4 cup (59.15 ml)	butter or margarine

(Recommendation: prepare a day in advance.)

1. Preheat oven to 350°F (176.6°C).
2. Drain mixed fruit and set aside.
3. Lightly grease an 8- or 9-inch (20.32- or 22.86-cm) baking pan. Place the mixed fruit on the bottom of the pan.
4. In a smaller bowl, combine all of the dry ingredients. Cut in the butter or margarine with a pastry blender. Sprinkle mixture over mixed-fruit filling.
5. Bake for 30 to 35 minutes in conventional oven to a minimum internal temperature of 135°F (57.2°C) for 15 seconds.
6. Cool properly. Cool hot food from 135°F to 70°F (57.2°C to 21.1°C) within 2 hours; you then have an additional 4 hours to go from 70°F to 41°F (21.1°C to 5°C) or lower for a maximum total cool time of 6 hours.
7. Store in refrigeration at 41°F (5°C) or lower.
8. Reheat 165°F (73.9°C) for 15 seconds within 2 hours, serve warm.

As you can see in this example, the recipe is detailed with **ingredients, required measurements, smallwares,** and **equipment.** Product instructions are just one prerequisite program required for HACCP in your foodservice operation that we will put to good use. This is discussed in further detail in Star Point 3, Hazard Analysis.

■ EQUIPMENT

The prerequisite program for commercial foodservice equipment is based on product instructions. In the example of the mixed-fruit crisp, it is evident that the major equipment needed is an oven, pastry blender, refrigerator, and hot-holding equipment. The smallwares equipment needed is a baking pan, mixing bowls, spoons, can opener, timer, and measuring tools. Everyone knows you can't do a job properly without the right tools.

The benefits of using the right equipment are food safety and efficiency, plus better quality of food that increases sales and productivity, thereby saving labor cost. Dirty equipment, no preventive care, or waiting to make repairs is not only unsafe but will cost you more money because the equipment is not working efficiently, potentially causing more damage and in some cases completely destroying the piece of equipment. Additionally, these oversights could put employees in harm's way, injuring them or worse. No matter the foodservice operation, when injury occurs, it has a direct impact by lowering team morale and decreasing productivity, resulting in a substantial financial loss.

The efficiency of your foodservice is based on purchasing and using the correct commercial foodservice equipment. It is not mandatory; however, the industry standard is for all equipment to have the NSF International/UL (Underwriters Laboratories) seals of approval. Here is a basic list of minimum equipment standards that are safe and sanitary:

★ Equipment should come with written specifications. The equipment specifications are normally required by regulatory officials for your plan reviews. These specifications also provide instructions on how to install, utility (electrical/gas) requirements, information on performance tests including maximum performance capability, and recommendations on equipment maintenance.

★ Equipment construction requires food-contact surfaces to be smooth, nonporous, corrosion-resistant, and nontoxic. All corners and edges must be rounded off. If coating materials are used, they must be USDA or FDA approved to resist chipping, be nontoxic, and be cleanable. Additional sanitary design factors to prevent bacteria buildup must be considered, such as overlapping parts, drainage, exposed threads, and crevices.

★ All equipment must be simple to disassemble and easy to clean and maintain. The key is to have all parts of equipment readily accessible for cleaning, sanitizing, maintenance, and inspection without the use of tools. Follow preventive maintenance programs and equipment calibration schedules. Always keep an inspection and equipment maintenance log to track the preventive maintenance care of your equipment.

The final equipment prerequisite program is to set guidelines for repairing or replacing equipment and smallwares that fall below standard. These guidelines are based on the manufacturer's recommendations to prevent microbial growth and the direction of your regulatory agency. Another resource you can use is a third-party auditing company to help in the safety evaluation. Most foodservice operators use employee feedback/complaints and the tallying of maintenance bills to help in the decision process of "Do I replace or do I repair the equipment?"

STAR KNOWLEDGE EXERCISE: EQUIPMENT

In the Star Knowledge Exercise below, use the Key to determine the typical use or function of the following pieces of equipment. Using the Key, write the letter next to the piece of equipment. The first one is done for you.

Key Typical Use

A. Receiving and storage

B. Preparation

C. Service area

D. Ware washing

E. Waste removal

Equipment	Typical Use
1. Trash compactor	E
2. Mixer	
3. Convection oven	
4. Pot sink	
5. Hot-food table	

Equipment	Typical Use
6. Broiler	
7. Beverage dispenser	
8. Bain-marie	
9. Steamer	
10. Slicers	
11. Garbage disposal	
12. Grill	
13. Ice machine	
14. Freezer	
15. Scales	
16. Pulper	
17. Microwave oven	
18. Meat saw	
19. Cart	
20. Ware-washing machine	
21. Walk-in refrigerator	
22. Steam-jacketed kettle	
23. Coffee urn	
24. Fryer	
25. Dish-washing machine	

■ FACILITY DESIGN

Sanitary facility design is the next step that will keep food safe as it travels through the operation. It is more cost-effective to do it right the first time. Take advantage of the situation if you have the luxury of designing your facility and equipment placement from a "blank slate" and with preplanning. Doing so avoids having to go back and fix any design flaws due to poor planning or cutting corners. In an existing facility, you have to work with what you have. But you need to examine the possibilities of adding, moving, and modifying the facility and the equipment. Regulatory agencies mandate and have specific laws pertaining to a sanitary facility, and in the 2005 FDA Model Food Code, it is one of the responsibilities of the person in charge to always operate a safe and sanitary foodservice facility.

Whether you are planning a foodservice facility from scratch or working with an existing footprint, the goal is to prevent the contamination of food, starting with the location. In business, a crucial factor for success is location, location, location. The location is usually considered first for real-estate value, traffic studies, and marketing potential, but that is not enough. The area surrounding your chosen location needs to be a factor in controlling sanitary facility design. What contamination can occur from the surrounding area? Are there any airborne contaminants that could affect your business? Odors? Are the structures infested with cockroaches? Do you have a pond with ducks and geese? Do your employees step in the duck and geese waste products and transport these harmful microorganisms into your facility? Is your facility near a river that rodents use to occasionally visit your operation for a bite to eat? These are examples of **environmental considerations** to apply to window/door placement, ventilation systems, pest control systems, waterproofing, drainage systems, shoe/boot cleaning systems, methods of food delivery, and so on.

Besides picking a good location, you must be sure the facility itself is correctly constructed with a proper sanitary facility design. Corporate chain stores and franchise organizations duplicate successful facility design over and over again by using similar sanitary design. Chains do this because they know their menu, equipment, and how to flow food through their facility so cross-contamination does not occur. Additionally, there is a huge savings of money on labor, creative design, architecture fees, and engineering costs associated with electrical and mechanical systems and plumbing.

Sanitary facility design is a prerequisite program because it is a proactive approach to **manage cross-contamination and prevent microbial growth.** The facility design needs to take into consideration the flow of the products through the operation to the customer. The design begins with the outside, then accounts for ingredients coming in, the proper storage of these products, complete preparation and processes used, as well as proper disposal of waste. Here is a list of facility considerations:

- ★ Interior materials (walls, floors, ceiling)
- ★ Equipment locations (flow)
- ★ Spacing of shelving and equipment 6 inches (15.23 cm) off of the floor and away from walls
- ★ Easy to clean and sanitize
- ★ Adequate lighting
- ★ Proper ventilation
- ★ Appropriate temperature: 50°F to 70°F (9.9°C to 21.1°C)
- ★ Correct humidity: 50%–60%
- ★ Potable water source
- ★ Water control (floor drains, self-draining equipment)
- ★ Effective plumbing (back-flow prevention devices, air gaps, vacuum breakers)

Whether you are dealing with new construction or modifying an existing structure, once your equipment is in place, the best practice going forward is to use a master cleaning schedule. The **master cleaning schedule** involves the who, what, where, when, why, and how of cleaning and sanitizing hoods, filters, grease traps, ceilings, walls, floors, and food-contact surfaces. This prerequisite is usually audited by visual inspections of the person in charge, who performs a "manager's walk" or uses a detailed checklist of your equipment and facility.

★ UNDERSTANDING FOOD SAFETY

■ USING STANDARD OPERATING PROCEDURES (SOPs)

Standard Operating Procedures (SOPs) are **required** for all HACCP plans. They provide the acceptable practices and procedures that your foodservice organization requires you to follow. SOPs are only effective if they are followed! We will now define standard operating procedures in detail and provide an example of one. It is important for you to understand that SOPs play a large role in your HACCP plan and the safety of your facility and the food served.

SOPs are the practices and procedures in food production that ensure we produce safe food. SOPs must be in writing, and SOPs are required for a HACCP plan. Also, SOPs keep our foodservice operation **consistent** and our customers **safe**. We recommend that you take advantage of Star Point 1 in developing and then applying these food safety standard operating procedures to your foodservice operation.

The U.S. Department of Agriculture (USDA, www.usda.gov) has organized the SOPs in a consistent format. This format includes a description of the purpose, scope, key words, instructions, monitoring, corrective action, verification, and record keeping. This format also indicates dates for implementation, review, and revision and requires a signature verifying each action has taken place. Following are two examples of food safety standard operating procedures for purchasing and receiving.

SOP: Purchasing (Sample)

Purpose: To prevent contamination of food and to ensure safe foods are served to customers by purchasing food products from approved suppliers. These suppliers must be approved by appropriate regulatory services.

Scope: This procedure applies to foodservice managers who purchase foods from approved suppliers.

Key Words: Approved suppliers, regulatory services

Instructions: Contact regulatory services to ensure you are purchasing foods from approved suppliers. To find out if a supplier is approved, call

- ★ CDC (Centers for Disease Control and Prevention) Food Safety Office—404-639-2213 or www.cdc.gov
- ★ EPA (Environmental Protection Agency)—202-272-0167 or www.epa.gov
- ★ FSIS (Food Safety and Inspection Service)—888-674-6854 or www.fsis.usda.gov
- ★ FDA (Food and Drug Administration)—888-463-6332 or www.cfsan.fda.gov

1.	Domestic/imported food (including produce, bottled water, and other foods) *but not meat and poultry*	★ Evidence of regulatory oversight: copy of suppliers, local enforcement agency permit, state or federal registration or license, or a copy of the last inspection report
		★ Third-party audit results [many vendors now provide third-party guarantees, including NSF International or American Institute of Baking (AIB)]
		★ Microbiological or chemical analysis/testing results
		★ Person-in-the-plant verification (i.e. chain food facilities may have their own inspector monitor food they buy)
		★ Self-certification (guarantee) by a wholesale processor based on HACCP
		★ For raw agricultural commodities such as produce, certification of Good Agricultural Practices or membership in a trade association such as the United Fresh Fruit and Vegetable Association
		★ A copy of a wholesale distributor or processor's agreement with its suppliers of food safety compliance
2.	Domestic/imported meat, poultry, and related products such as meat- or poultry-containing stews, frozen foods, and pizzas	★ USDA mark on meat or poultry products
		★ Registration of importers with USDA
3.	Fish and fish products	★ Evidence of regulatory oversight: copy of suppliers' local enforcement agency permit, state or federal registration or license, or a copy of the last inspection report
		★ Third-party audit results
		★ Person-in-the-plant verification
		★ Self-certification (guarantee) by a wholesale processor based on HACCP
		★ A copy of a wholesale distributor or processor's agreement with its suppliers of HACCP compliance
		★ U.S. Department of Commerce (USDC) approved list of fish establishments and products located at seafood.nmfs.noaa.gov
4.	Shellfish	★ Shellfish tags
		★ Listing in current *Interstate Certified Shellfish Shippers* publication
		★ Gulf oyster treatment process verification if sold between April 1 and October 31 (November 1 to March 31 certification may be used in lieu of warning signs)
		★ USDC approved list of fish establishments and products located at seafood.nmfs.noaa.gov

5.	Drinking water (nonbottled water)	★ A recent certified laboratory report demonstrating compliance with drinking water standards
		★ A copy of the latest inspection report
6.	Alcoholic beverages	★ Third-party audit results
		★ Self-certification (guarantee) by a wholesale processor based on HACCP
		★ Person-in-the-plant verification
		★ Evidence of regulatory oversight: copy of suppliers' local enforcement agency permit, state or federal registration or license, or a copy of the last inspection report
		★ A copy of a wholesale distributor or processor's agreement with its suppliers of food safety compliance

Monitoring:

1. Inspect invoices or other documents to determine approval by a regulatory agency.
2. Foodservice managers should be encouraged to make frequent inspections of the suppliers' on-site facilities, manufacturing facilities, and processing plants/farms. Inspections determine cleanliness standards and ensure that HACCP plans are in place.

Corrective Action:

Foodservice purchasing managers must find a new supplier if the supplier is not approved by the above regulatory services.

Verification and Record Keeping:

The foodservice purchasing manager will maintain all documentation from food suppliers. Documentation must be maintained for three years plus the current year.

Date Implemented: _____	By: _____
Date Reviewed: _____	By: _____
Date Revised: _____	By: _____

SOP: Receiving Deliveries (Sample)

Purpose: To ensure that all food is received fresh and safe when it enters the foodservice operation, and to transfer food to proper storage as quickly as possible

Scope: This procedure applies to foodservice employees who handle, prepare, or serve food.

Key Words: Cross-contamination, temperatures, receiving, holding, frozen goods, delivery

Instructions:

1. Train foodservice employees who accept deliveries on proper receiving procedures.
2. Schedule deliveries to arrive at designated times during operational hours.
3. Post the delivery schedule, including the names of vendors, days and times of deliveries, and drivers' names.
4. Establish a rejection policy to ensure accurate, timely, consistent, and effective refusal and return of rejected goods.
5. Organize freezer and refrigeration space, loading docks, and storerooms before receiving deliveries.
6. Before deliveries, gather product specification lists and purchase orders, temperature logs, calibrated thermometers, pens, and flashlights, and be sure to use clean loading carts.
7. Keep receiving area clean and well lighted.
8. Do not touch ready-to-eat foods with bare hands.
9. Determine whether foods will be marked with the date of arrival or the "use by" date and mark accordingly upon receipt.
10. Compare delivery invoice against products ordered and products delivered.
11. Transfer foods to their appropriate locations as quickly as possible.

Monitoring:

1. Inspect the delivery truck when it arrives to ensure that it is clean, free of putrid odors, and organized to prevent cross-contamination. Be sure refrigerated foods are delivered on a refrigerated truck.
2. Check the interior temperature of refrigerated trucks.
3. Confirm vendor name, day and time of delivery, as well as driver's identification before accepting delivery. If the driver's name is different than what is indicated on the delivery schedule, contact the vendor immediately.
4. Check frozen foods to ensure that they are all frozen solid and show no signs of thawing and refreezing, such as the presence of large ice crystals or liquids on the bottom of cartons.

5. Check the temperature of refrigerated foods.

   ★ For fresh meat, fish, dairy, and poultry products, insert a clean and sanitized thermometer into the center of the product to ensure a temperature of 41°F (5°C) or below.

   ★ For packaged products, insert a food thermometer between two packages, being careful not to puncture the wrapper. If the temperature exceeds 41°F (5°C), it may be necessary to take the internal temperature before accepting the product.

   ★ For eggs, the interior temperature of the truck should be 45°F (7.2°C) or below.

6. Check dates of milk, eggs, and other perishable goods to ensure safety and quality.

7. Check the integrity of food packaging.

8. Check the cleanliness of crates and other shipping containers before accepting products. Reject foods that are shipped in dirty crates.

Corrective Action:

Reject the following:

★ Frozen foods with signs of previous thawing

★ Cans that have signs of deterioration—swollen sides or ends, flawed seals or seams, dents, or rust

★ Punctured packages

★ Expired foods

★ Foods that are out of the safe temperature zone or deemed unacceptable by the established rejection policy

Verification and Record Keeping:

The designated team member needs to record temperatures and corrective actions taken on the delivery invoice or on the receiving log. The foodservice manager will verify that foodservice employees are receiving products using the proper procedure by visually monitoring receiving practices during the shift and reviewing the receiving log at the close of each day. Receiving and corrective action logs are kept on file for a minimum of 1 year.

Date Implemented: _____	By: _____
Date Reviewed: _____	By: _____
Date Revised: _____	By: _____

STAR KNOWLEDGE EXERCISE: STORAGE SOP

In the space below, list the directions for instructions, monitoring, corrective action, verification, and record keeping needed for proper storage.

SOP: Storage (Exercise)

Purpose: To ensure that food is stored safely and put away as quickly as possible after it enters the foodservice operation

Scope: This procedure applies to foodservice employees who handle, prepare, or serve food.

Key Words: Cross-contamination, temperatures, storing, dry storage, refrigeration, freezer

Instructions:

1.

2.

3.

4.

5.

6.

7.

8.

9.

10.

Monitoring:

Corrective Action:

Verification and Record Keeping:

Date Implemented: _____ By: _____

Date Reviewed: _____ By: _____

Date Revised: _____ By: _____

★ COMMON FOODBORNE ILLNESSES

This section helps you understand how to manage and control the microorganisms that cause foodborne illness. Vomiting, diarrhea, stomach cramps, and flulike symptoms are the most common symptoms associated with foodborne illnesses.

Ask yourself these questions:

★ Have you ever eaten food that made you sick?

★ Did you vomit?

★ Did you have stomach cramps?

★ Did you have diarrhea?

★ Did you cough up worms?

These symptoms may be the result of a foodservice facility not following prerequisite programs such as standard operating procedures. This chapter should help you to understand food safety so that you can protect yourself, your family, your friends, your neighbors, your fellow team members, your facility, and most of all, your customers.

The people at the **most risk** for foodborne Illness are

★ Children

★ People who are already sick

★ People taking medication

★ Pregnant women

★ Elderly people

Courtesy PhotoDisc, Inc.

If SOPs are not followed, you and your customers may contract a foodborne illness. Microorganisms are found everywhere, including the foods that you eat. Microorganisms can be categorized as bacteria, viruses, parasites, and fungi. Some types of microorganisms are helpful in the production of certain medicines (penicillin and vaccines) and foods (cheese, beer, and bread). Some fungi (yeast) are used in producing beer and making bread rise. Also, some bacteria help in our digestion process.

However, some microorganisms are **pathogenic**, which means they are disease-causing microorganisms. Every day we eat pathogens, but most people have enough antibodies to fight off a certain amount of the harmful aspects of pathogens when they are ingested. However, if there are too many pathogens in your system and your antibodies can't fight them off, you will become ill. Illnesses that travel to you through food are called **foodborne illnesses**. A foodborne illness will occur when pathogenic microorganisms are carried into your system through the food that you eat. This is referred to "contracting a foodborne illness." A foodborne illness outbreak may occur when as few as two people consume the same food and get the same illness.

Bacteria are managed through food, acidity, time, temperature, oxygen, and moisture, or the acronym **FATTOM**. These six elements are the conditions needed for microorganisms to grow. Let's look at each condition more closely:

- ★ **F**—Is the **food** potentially hazardous, or is time/temperature control for safety of food needed?
- ★ **A**—What is the **acidity**? How does the acidity interact with the water activity?
- ★ **T**—What is the total amount of **time** in the temperature danger zone (TDZ)?
- ★ **T**—Is the food in the **temperature** danger zone of 41°F to 135°F (5°C to 57.2°C)?
- ★ **O**—Is a reduced **oxygen** method used to store the food?
- ★ **M**—What is the **moisture** (water activity)? How does the water activity interact with the acidity?

Viruses can be controlled by hand washing, maintaining proper personal hygiene, cleaning, and sanitizing. **Parasites** are destroyed by cooking food to the minimum internal cooking temperature or by proper freezing methods. **Fungi** are controlled by purchasing from a reputable supplier, conducting a visual inspection, and carefully monitoring time and temperature.

Listed in the following are the most common foodborne illnesses and the sources associated with each illness. You need to be aware of the biological hazards related to the foods that you produce, serve, or eat. In addition, this information will be useful in HACCP Principle 1, "Conducting a Hazard Analysis."

■ VIRUSES

Disease: Virus	Typically a result of
Hepatitis A	★ Not washing hands properly; infected employee still coming to work; receiving shellfish from unapproved sources; handling RTE foods, water, and ice with contaminated hands ★ Highly contagious—**must report** to person in charge
Norovirus	★ Poor personal hygiene; receiving shellfish from unapproved sources; and using unsanitary/nonchlorinated water ★ Very common with people in close quarters for long periods of time (dormitories, offices, and cruise ships) ★ Highly contagious—**must report** to person in charge
Rotavirus Gastroenteritis	★ Not cooking foods to the required minimum internal temperature; not maintaining time/temperature control ★ Poor personal hygiene ★ Not cleaning and sanitizing properly

■ BACTERIA

Disease: Bacteria	Typically a result of
Salmonellosis (salmonella)	★ Improper handling and cooking of eggs, poultry, and meat; contaminated raw fruits and vegetables ★ Highly contagious—**must report** to person in charge
Shigellosis (bacillary dysentery)	★ Flies, water, and foods contaminated with fecal matter ★ Improperly handling ready-to-eat foods and time/temperature abuse ★ Found in the intestine of humans. Wash your hands! ★ Highly contagious—**must report** to person in charge

Disease: Bacteria	Typically a result of
Hemorrhagic colitis (E. coli)	★ Undercooked ground beef, unpasteurized juice/cider and dairy products, contact with infected animals, and cross-contamination ★ Highly contagious—**must report** to person in charge
Bacillus cereus	★ Improper holding, cooling, and reheating rice products, potatoes, and starchy foods
Botulism	★ Time and temperature abuse, garlic-and-oil mixtures, improperly sautéing and holding sautéed onions, serving home-canned products and improperly cooling leftovers, improper processing and storing of canned goods
Campylobacteriosis	★ Not cooking food, especially chicken, to proper internal temperatures, cross-contamination and using unpasteurized milk and untreated water.
Clostridium perfringens	★ Improper temperature control, reheating, cooling, and holding cooked food like meat, poultry, beans, and gravy ★ Found in the intestinal tract of humans
Listeriosis	★ Not cooking food to the required minimum internal temperature, not washing raw vegetables, and not cleaning/sanitizing food preparation surfaces ★ Associated with hot dogs, processed lunch meats, soft cheeses, unpasteurized milk/dairy products, and cross-contamination during packaging and processing
Staphylococcal gastroenteritis	★ Unwashed bare hands, having a skin infection while handling and preparing food; found on skin, hair, nose, mouth, and throat ★ Improperly refrigerating or cooling prepared food
Vibrio	★ Eating raw or partially cooked crabs, clams, shrimp, and oysters, receiving seafood from an unapproved supplier
Yersiniosis	★ Using unsanitary/nonchlorinated water and cross-contamination, unpasteurized milk and not thoroughly cooking food to the required minimum internal temperature

■ PARASITES

Parasites need a host to survive. A parasitic host can be humans, rats, pigs, bears, walruses, fish, and wild game.

Disease: Parasites	Typically a result of
Anisakiasis	★ Receiving seafood from unapproved sources, and serving undercooked or raw seafood
Cyclosporiasis	★ Drinking or using unsanitary water supplies
Giardiasis	★ Drinking or using unsanitary water supplies
Intestinal Cryptosporidiosis	★ Drinking or using unsanitary water supplies
Trichinosis	★ Improperly cooking pork and game meat, improperly cleaning and sanitizing equipment and utensils used to process pork and other meats, receiving meats from an unapproved supplier
Toxoplasmosis	★ Not properly washing hands after touching raw vegetables, cat feces, soil, or raw/undercooked meats (particularly poultry, lamb or wild game); and not cooking meats to the required minimum internal temperature

■ DUTY TO REPORT FOODBORNE ILLNESS DISEASES

If you find out you have contracted or have been exposed to any of the illnesses in the bacteria, virus, and parasite tables, notify your supervisor immediately. Your employees also need to report these illnesses to you or another person in charge.

Five of these foodborne illnesses are considered **highly contagious**. They are known as the **Big 5**. The newest addition to this list is the norovirus, which was added to the list in 2005. The CDC estimates that norovirus is the leading cause of foodborne illness in the United States. Following is the complete list of the Big 5:

★ **Norovirus** (also known as "Norwalk-like virus," "small round-structured virus," and "winter vomiting disease")

★ **Salmonellosis** (also known as typhoid fever)

★ **Shigellosis** (also known as dysentery)

★ **Hemorrhagic colitis** (also known as E. coli)

★ **Hepatitis A** (noticed by a jaundiced condition)

Some companies have an infectious disease policy and procedures to follow. According to the 2005 FDA Model Food Code (www.cfsan.fda.gov), Forms 1-A, 1-B, and 1-C are designed to assist those responsible for reporting foodborne diseases. The 2005 FDA Model Food Code specifies that the **permit holder is responsible** for requiring applicants to use (Form 1-A) and food employees to use (Form 1-B) to report certain symptoms, diagnoses, past illnesses, high-risk conditions, and foreign travel, as they relate to diseases transmitted through food by infected workers. The **food employee is personally responsible** for reporting this information to the person in charge.

Here is another opportunity to take action and use these forms to make a difference in your operation. Once an employee is confirmed with one of the Big 5 illnesses, the health practitioner or physician needs to complete Form 1-C. If the health practitioner denies your employee to return to work, because the employee might still be a carrier, then your employee might want to consult an infectious disease specialist. **FDA forms 1-A, 1-B, and 1-C follow (on pages 22 to 27) and can also be found at www.cfsan.fda.gov/~acrobat/fc05-a7.pdf.**

■ MAJOR FOOD ALLERGENS

Some of the symptoms associated with a foodborne illness are the same symptoms associated with an allergic reaction. When it comes to food safety, allergies are just as dangerous as foodborne illnesses. In the 2005 FDA Model Food Code, recent studies indicate that over 11 million Americans suffer from one or more food allergies. A **food allergy** is caused by a naturally occurring protein in a food or a food ingredient, which is referred to as an **allergen**.

Is your customer having an allergic reaction to food? Let's find out . . .

★ Is your customer's throat getting tight?

★ Does your customer have shortness of breath?

★ Does your customer have itching around the mouth?

★ Does your customer have hives?

Anyone can be allergic to **anything**. Sometimes people don't know they have a food allergy until they have a reaction to a food that causes some or all of the symptoms listed. In severe cases, anaphylactic shock and death may result. We have included this in the food safety standard operating procedures point of the HACCP Star because allergies are a growing concern in the effort to serve safe food.

If you are someone who has had an allergic reaction to food, you can understand how important it is to know what is in the foods you, your family, friends, neighbors, fellow team members, and customers are consuming.

The first step is to be aware of the most common allergens. Although there are others, the most common are known as **major food allergens** and the **Big 8.** These foods account for 90 percent or more of all food allergies. They are as follows:

★ Shellfish (crab, lobster, or shrimp)

★ Fish (bass, flounder, or cod)

★ Peanuts

★ Tree nuts (almonds, pecans, chestnuts, pistachios, Brazil nuts, etc.)

★ Milk

★ Eggs

★ Soy/tofu

★ Wheat

(continued on page 28)

FORM 1-A	**Conditional Employee and Food Employee Interview**
	Preventing Transmission of Diseases through Food by Infected Food Employees or Conditional Employees with Emphasis on illness due to Norovirus, *Salmonella* **Typhi**, *Shigella* spp., Enterohemorrhagic (EHEC) or Shiga toxin-producing *Escherichia coli* (STEC), or hepatitis A Virus

The purpose of this interview is to inform conditional employees and food employees to advise the person in charge of past and current conditions described so that the person in charge can take appropriate steps to preclude the transmission of foodborne illness.

Conditional employee name (print) _____

Food employee name (print) _____

Address _____

Telephone *Daytime:*_____ *Evening:* _____

Date _____

Are you suffering from any of the following symptoms? (Circle one)

		If YES, Date <u>of Onset</u>
Diarrhea?	YES / NO	_____
Vomiting?	YES / NO	_____
Jaundice?	YES / NO	_____
Sore throat with fever?	YES / NO	_____

Or

Infected cut or wound that is open and draining, or lesions containing pus on the hand, wrist, an exposed body part, or other body part and the cut, wound, or lesion not properly covered?   YES / NO

(Examples: *boils and infected wounds, however small*)

<u>In the Past:</u>

Have you ever been diagnosed as being ill with typhoid fever (*Salmonella* Typhi)   YES / NO
If you have, what was the date of the diagnosis? _____
If within the past 3 months, did you take antibiotics for *S. Typhi?*   YES / NO
    If so, how many days did you take the antibiotics? _____
    If you took antibiotics, did you finish the prescription? _____   YES / NO

<u>History of Exposure:</u>

1. **Have you been suspected of causing or have you been exposed to a confirmed foodborne disease outbreak recently?**   YES / NO
      If YES, date of outbreak: _____
a. **If YES, what was the cause of the illness and did it meet the following criteria?**
    Cause: _____
    i. **Norovirus (last exposure within the past 48 hours)**   Date of illness outbreak _____
    ii. *E. coli* **O157:H7 infection (last exposure within the past 3 days)**   Date of illness outbreak _____
    iii. **Hepatitis A virus (last exposure within the past 30 days)**   Date of illness outbreak _____
    iv. **Typhoid fever (last exposure within the past 14 days)**   Date of illness outbreak _____
    v. **Shigellosis (last exposure within the past 3 days)**   Date of illness outbreak _____

FORM 1-A (continued)

b. **If YES, did you:**
    i.   **Consume food implicated in the outbreak?** _____
    ii. **Work in a food establishment that was the source of the outbreak?** _____
    iii. **Consume food at an event that was prepared by person who is ill?** _____

2. **Did you attend an event or work in a setting, recently where there was a confirmed disease outbreak?**   **YES / NO**

     If so, what was the cause of the confirmed disease outbreak? _____

     If the cause was one of the following five pathogens, did exposure to the pathogen meet the following criteria?

     a. **Norovirus (last exposure within the past 48 hours)**   **YES / NO**
     b. **E. coli O157:H7 (or other** EHEC/STEC **(last exposure within the past 3 days)**   **YES / NO**
     c. **Shigella spp. (last exposure within the past 3 days)**   **YES / NO**
     d. **S. Typhi (last exposure within the past 14 days)**   **YES / NO**
     e. **hepatitis A virus (last exposure within the past 30 days)**   **YES / NO**

     Do you live in the same household as a person diagnosed with Norovirus, Shigellosis, typhoid fever, hepatitis A, or illness due to E. coli O157:H7 or other EHEC/STEC**?**
                                                      YES / NO Date of onset of illness _____

3. **Do you have a household member attending or working in a setting where there is a confirmed disease outbreak of Norovirus, typhoid fever, Shigellosis, EHEC/STEC infection, or hepatitis A?**
                                                      YES / NO Date of onset of illness _____

     Name, Address, and Telephone Number of your Health Practitioner or doctor:
     Name _____
     Address _____
     Telephone – *Daytime:* _____ *Evening:* _____

Signature of Conditional Employee _____ **Date** _____

Signature of Food Employee _____ **Date** _____

Signature of Permit Holder or Representative _____ **Date** _____

FORM 1-B	**Conditional Employee or Food Employee Reporting Agreement**
	Preventing Transmission of Diseases through Food by Infected Conditional Employees or Food Employees with Emphasis on illness due to Norovirus, *Salmonella* Typhi, *Shigella* spp., Enterohemorrhagic (EHEC) or Shiga toxin-producing *Escherichia coli* (STEC), or hepatitis A Virus

The purpose of this agreement is to inform conditional employees or food employees of their responsibility to notify the person in charge when they experience any of the conditions listed so that the person in charge can take appropriate steps to preclude the transmission of foodborne illness.

I AGREE TO REPORT TO THE PERSON IN CHARGE:

Any Onset of the Following Symptoms, Either While at Work or Outside of Work, Including the Date of Onset:

1. Diarrhea
2. Vomiting
3. Jaundice
4. Sore throat with fever
5. Infected cuts or wounds, or lesions containing pus on the hand, wrist , an exposed body part, or other body part and the cuts, wounds, or lesions are not properly covered (*such as boils and infected wounds, however small*)

Future Medical Diagnosis:

Whenever diagnosed as being ill with Norovirus, typhoid fever (*Salmonella* Typhi), shigellosis (*Shigella* spp. infection), *Escherichia* coli O157:H7 or other EHEC/STEC infection, or hepatitis A (hepatitis A virus infection)

Future Exposure to Foodborne Pathogens:

1. Exposure to or suspicion of causing any confirmed disease outbreak of Norovirus, typhoid fever, shigellosis, *E.* coli O157:H7 or other EHEC/STEC **infection, or hepatitis A.**
2. A household member diagnosed with Norovirus, typhoid fever, shigellosis, illness due to EHEC/STEC, **or hepatitis A.**
3. A household member attending or working in a setting experiencing a confirmed disease outbreak of Norovirus, typhoid fever, shigellosis, *E.* coli O157:H7 or other EHEC/STEC **infection, or hepatitis A.**

I have read (or had explained to me) and understand the requirements concerning my responsibilities under the **Food Code** and this agreement to comply with:

1. Reporting requirements specified above involving symptoms, diagnoses, and exposure specified;
2. Work restrictions or exclusions that are imposed upon me; and
3. Good hygienic practices.

I understand that failure to comply with the terms of this agreement could lead to action by the food establishment or the food regulatory authority that may jeopardize my employment and may involve legal action against me.

Conditional Employee Name (please print) _____

Signature of Conditional Employee _____ **Date** _____

Food Employee Name (please print) _____

Signature of Food Employee _____ **Date** _____

Signature of Permit Holder or Representative _____ **Date** _____

FORM 1-C	**Conditional Employee or Food Employee Medical Referral**

Preventing Transmission of Diseases through Food by Infected Food Employees with Emphasis on Illness due to Norovirus, Typhoid fever (*Salmonella* Typhi), **Shigellosis** (*Shigella* spp.), *Escherichia coli* **O157:H7** or other Enterohemorrhagic (EHEC) or Shiga toxin-producing *Escherichia* **coli** (STEC), and hepatitis A Virus

The **Food Code** specifies, under *Part 2-2 Employee Health Subpart 2-201 Disease or Medical Condition*, that Conditional Employees and Food Employees obtain medical clearance from a health practitioner licensed to practice medicine, unless the Food Employees have complied with the provisions specified as an alternative to providing medical documentation, whenever the individual:

1. Is chronically suffering from a symptom such as **diarrhea;** *or*
2. Has a **current illness** involving Norovirus, typhoid fever (*Salmonella* **Typhi**), shigellosis (*Shigella* spp.) *E. coli* **O157:H7** infection (or other EHEC/STEC), or hepatitis A virus (hepatitis A), *or*
3. Reports *past illness* involving typhoid fever (*S.* **Typhi**) within the past three months (while salmonellosis is fairly common in U.S., typhoid fever, caused by infection with *S.* **Typhi**, is rare).

Conditional employee being referred: (Name, please print) _____

Food Employee being referred: (Name, please print) _____

4. Is the employee assigned to a food establishment that serves a population that meets the Food Code definition of a **highly susceptible population** such as a day care center with preschool age children, a hospital kitchen with immunocompromised persons, or an assisted living facility or nursing home with older adults? **YES** ☐   **NO** ☐

Reason for Medical Referral: The reason for this referral is checked below:
☐ Is chronically suffering from vomiting or diarrhea; or (specify) _____
☐ Diagnosed or suspected Norovirus, typhoid fever, shigellosis, *E. coli* O157:H7 (or other EHEC/STEC) infection, or hepatitis A. (Specify) _____
☐ Reported past illness from typhoid fever within the past 3 months. (Date of illness) _____
☐ Other medical condition of concern per the following description: _____

Health Practitioner's Conclusion: (Circle the appropriate one; refer to reverse side of form)
☐ Food employee is free of **Norovirus** infection, typhoid fever (*S.* **Typhi** infection), *Shigella* spp. infection, *E. coli* O157:H7 (or other **EHEC/STEC** infection), or **hepatitis A** virus infection, and may work as a food employee without restrictions.
☐ Food employee is an asymptomatic shedder of *E.* coli O157:H7 (or other **EHEC/STEC**), *Shigella* spp., or Norovirus, and is restricted from working with exposed food; clean equipment, utensils, and linens; and unwrapped single-service and single-use articles in food establishments that do not serve highly susceptible populations.
☐ Food employee is not ill but continues as an asymptomatic shedder of *E. coli* O157:H7 (or other **EHEC/STEC**), *Shigella* spp. and should be excluded from food establishments that serve highly susceptible populations such as those who are preschool age, immunocompromised, or older adults and in a facility that provides preschool custodial care, health care, or assisted living.
☐ Food employee is an asymptomatic shedder of **hepatitis A** virus and should be excluded from working in a food establishment until medically cleared.
☐ Food employee is an asymptomatic shedder of **Norovirus** and should be excluded from working in a food establishment until medically cleared, or for at least 24 hours from the date of the diagnosis.
☐ Food employee is suffering from Norovirus, typhoid fever, shigellosis, *E. coli* O157:H7 (or other **EHEC/STEC** infection), or **hepatitis A** and should be excluded from working in a food establishment.

FORM 1-C (continued)

COMMENTS: (In accordance with Title I of the Americans with Disabilities Act (ADA) and to provide only the information necessary to assist the food establishment operator in preventing foodborne disease transmission, please confine comments to explaining your conclusion and estimating when the employee may be reinstated.)

_____

_____

_____

_____

_____

_____

_____

_____

Signature of Health Practitioner _____ **Date** _____

Paraphrased from the FDA Food Code for Health Practitioner's Reference

From Subparagraph 2-201.11(A)(2)   Organisms of Concern:

Any foodborne pathogen, with special emphasis on these 5 organisms:
   1. **Norovirus** 2. *S. Typhi* 3. *Shigella* spp. 4. *E. coli* O157:H7 (or other EHEC/STEC) 5. **Hepatitis A** virus

From Subparagraph 2-201.11(A)(1)   Symptoms:

Have any of the following symptoms:
   Diarrhea   **Vomiting**   **Jaundice**   **Sore throat with fever**

From Subparagraph 2-201.11(A)(4)-(5)   Conditions of Exposure of Concern:

   (1) Suspected of causing a foodborne outbreak or being exposed to an outbreak caused by 1 of the 5 organisms above, at an event such as a family meal, church supper, or festival because the person:
   Prepared or consumed an implicated food; or
   Consumed food prepared by a person who is infected or ill with the organism that caused the outbreak or who is suspected of being a carrier;
   (2) Lives with, and has knowledge about, a person who is diagnosed with illness caused by 1 of the 5 organisms; or
   (3) Lives with, and has knowledge about, a person who works where there is an outbreak caused by 1 of the 5 organisms.

From Subparagraph 2-201.12   Exclusion and Restriction:

Decisions to exclude or restrict a food employee are made considering the available evidence about the person's role in actual or potential foodborne illness transmission. Evidence includes:

   Symptoms   **Diagnosis**   **Past illnesses**   **Stool/blood tests**

In facilities serving highly susceptible populations such as day care centers and health care facilities, a person for whom there is evidence of foodborne illness is almost always <u>excluded</u> from the food establishment.

In other establishments such as restaurants and retail food stored, that offer food to typically healthy consumers, a person might only be <u>restricted</u> from certain duties, based on the evidence of foodborne illness.

Exclusion from any food establishment is required when the person is:
   • Exhibiting or reporting diarrhea or vomiting;
   • Diagnosed with illness caused by *S.* Typhi; or
   • Jaundiced within the last 7 days.

For *Shigella* spp. or *Escherichia coli* O157:H7 or other EHEC/STEC infections, the person's stools must be negative for 2 consecutive cultures taken no earlier than 48 hours after antibiotics are discontinued, and at least 24 hours apart or the infected individual must have resolution of symptoms for more than 7 days or at least 7 days have passed since the employee was diagnosed.

In August 2004, the Food Allergen Labeling and Consumer Protection Act was enacted, which defines the term "major food allergen." This definition was adopted in the 2005 FDA Model Food Code. As of January 1, 2006, the new law requires food manufacturers to identify in plain language on the label of food any major food allergen used as an ingredient. Also, the FDA is to conduct inspections to ensure food facilities comply with practices to reduce or eliminate cross contact of a food with any major food allergens that are not intentional ingredients of the food. This new law will help foodservice operators assist their customers by identifying allergens quicker and faster because the label will be easier to read.

Be aware that some allergy symptoms are actually a result of **intolerance** to certain foods. For example, milk has a type of sugar in it called lactose that inhibits digestion in some people, resulting in lactose intolerance. Consuming any dairy product may result in experiencing symptoms like nausea, diarrhea, abdominal bloating, excessive gas, and cramping.

Other allergens associated with food preparation are

- ★ **MSG,** or **monosodium glutamate** (used as a food additive/flavor enhancer).
- ★ **Sulfites** or **sulfur dioxide** (used as a vegetable freshener/potato whitening agent).
- ★ **Latex** (latex residue can be transferred from the latex gloves to foods, such as tomatoes, before they are served). It is recommended that employees not wear latex gloves when touching food.

Considering some allergens cause reactions that may be mild or may be severe enough to cause death, you should take the following steps to ensure your customers avoid eating foods to which they are allergic. First of all, think about how you would like to be treated if you were the customer with a food allergy. Although all customers are special, an allergic customer will have a limited amount of restaurants that he or she will be able to frequent. If we earn the trust of an allergic customer, we also earn their repeat business. Then consider the following steps:

1. **Ask the customer** if he or she has any food allergies.
2. **Know your company's SOP.** What should you do if your customer indicates he or she has a food allergy?
3. **Know your menu.** Describe *all* ingredients and the preparation of foods you are serving to anyone who asks, even if it is a "secret recipe."
4. **Be honest.** It is OK to say, "I don't know." Immediately ask your manager to assist you.
5. **Be careful.** Make sure your customer is not allergic to anything in the food you are serving. You should also make certain that he or she is not allergic to anything with which the food has come into contact (SOP: Prevent Cross-Contamination).
6. **Be thoughtful and concerned**, but never tell a customer you are sorry he or she has an allergy to certain foods, because no one is at fault for someone having an allergy.
7. Manage allergens by **limiting the contact** of food for any allergic customer. It is best if **only 1 person** handles the customer's entire food preparation and service. Even utensils and plates can cause cross-contamination of allergens to several surfaces.

★ STAR KNOWLEDGE EXERCISE: FOODBORNE ILLNESSES AND ALLERGENS

1. If you have been diagnosed with or come in contact with someone who has hepatitis A, what should you do?

2. Name the Big 5 foodborne illnesses. What should you do if you contract one of them?

3. How would you handle a customer who tells you he is allergic to walnuts but wants to order the chicken salad, which has toasted nuts in it?

★ INTERNATIONAL FOOD SAFETY ICONS

We all know what the blue handicapped parking sign means when we drive around a parking lot. Signs with simple pictures tell us when it is safe to cross the street, when to check the oil in our cars, or how to get to the airport. With the same purpose in mind, **International Food Safety Icons** help make food safety easier for everyone to understand and help you remember basic food safety rules and procedures for food preparation. Throughout this section, you will see the various International Food Safety Icons, which will help you succeed in becoming a HACCP Superstar. The International Food Safety Icons provide a visual definition and reminder of the standard operating procedures for the foodservice industry. The leadership team at a particular foodservice operation has the responsibility to establish policies, procedures, and recipes that must be followed. The International Food Safety Icons make it easy for everyone working at the establishment to **understand**, **remember**, and **reinforce** these procedures.

The following Food Safety Match Game gives you an overview of the standard operating procedures used in most foodservices. Check your knowledge of food safety by matching the International Food Safety Icons with the associated rule. Earn a point for each correct answer. Select the food safety rule that best fits the food safety symbol.

★ FOOD SAFETY MATCH GAME

 1. _____

 2. _____

 3. _____

A. Potentially Hazardous Foods—Time/Temperature Control for the Safety of Food (PHF/TCS)

B. Wash, Rinse, Sanitize.

C. Cooling Food.

D. Temperature Danger Zone (TDZ)—41°F to 135°F (5°C to 57.2°C).

E. Cook All Foods Thoroughly.

F. Cold Holding—Hold cold foods below 41°F (5°C).

H. Hot Holding—Hold hot foods above 135°F (57.2°C).

I. Do Not Cross-Contaminate—From raw to ready-to-eat or cooked foods.

J. Wash Your Hands.

K. No Bare-Hand Contact—Don't handle food with bare hands.

L. Do Not Work If Ill.

 4. _____

 5. _____

 6. _____

 7. _____

 8. _____

 9. _____

 10. _____

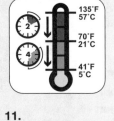 11. _____

How many points did you earn? _____

If you scored 10–11 points—Congratulations! You are a Food Safety Superstar!

If you scored 8–9 points—Good job! You have a basic understanding of food safety.

If you scored 5–7 points—The time for review is now! What a great opportunity to fine-tune your food safety skills.

If you scored 0–4 points—Everyone needs to start somewhere!

We are confident that if you follow the prerequisite programs, specifically the standard operation procedures in this manual, you will better understand why the basics of food safety must be mastered. The HACCP team must also master food safety basics as the first step toward creating an effective HACCP plan. You must know the proper ways to cook and prepare food before you can determine the mistakes being made in preparing the food at your facility. Then you need to write down the information in the form of a standard operating procedure to avoid mistakes in handling, preparing, and serving food to ensure food safety.

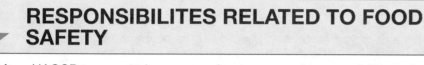

RESPONSIBILITES RELATED TO FOOD SAFETY

As a HACCP team member, some of your personal responsibilities related to providing safe food are staying home when sick, washing your hands, using gloves properly, and following a food-safe dress code. As we mentioned previously, each of the International Food Safety Icons represents a food safety standard operating procedure. Let's look at each one.

■ DO NOT WORK IF ILL

If you have gastrointestinal symptoms like running a **fever**, **vomiting**, and **diarrhea**, or **you are sneezing and coughing**, you should not work around or near food and beverages. If you are diagnosed with a foodborne illness, it is critical that you stay home until your physician gives you permission to work around food again. It is important for you to notify the person in charge/manager of any illnesses you may have, especially if you have norovirus, hepatitis A, E. coli, salmonellosis, or shigellosis, because these diseases must be reported to the regulatory authority.

According to the 2005 FDA Model Food Code (www.cfsan.fda.gov), the following chart summarizes the CDC list and compares the common symptoms of each pathogen. Symptoms may include diarrhea, fever, vomiting, jaundice, and sore throat with fever. CDC has no evidence that the HIV virus is transmissible via food. Therefore, a food employee who is HIV positive is not of concern unless he or she is suffering from a secondary illness listed in the chart. Lists I and II include pathogens likely to occur in foods.

| KEY: D = Diarrhea | F = Fever | V = Vomiting | J = Jaundice | S = Sore throat with fever |

LIST I. Pathogens often transmitted by food contaminated by infected persons who handle food.

	D	F	V	J	S
1. Noroviruses	D	F	V		
2. Hepatitis A virus		F		J	
3. *Salmonella* typhi		F			
4. *Shigella* species	D	F	V		
5. *Staphylococcus aureus*	D		V		
6. Streptococcus pyogenes		F			S

LIST II. Pathogens occasionally transmitted by food contaminated by infected persons who handle food, but usually transmitted by a contaminated source or in food processing or by non-foodborne routes.

	D	F	V	J	S
1. *Campylobacter jejuni*	D	F	V		
2. *Cryptosporidium parvum*	D				
3. Enterohemorrhagic *Escherichia coli*	D				
4. Enterotoxigen *Escherichia coli*	D		V		
5. *Giardia lamblia*	D				
6. Nontyphoidal *salmonella*	D	F	V		
7. *Vibrio cholerae* 01	D		V		
8. *Yersinia enterocolitica*	D	F	V		

- Hand antiseptics should only be used on clean hands.

- Hand antiseptics are not a substitute for hand washing.

- Use only antiseptics approved by the FDA.

■ WASH YOUR HANDS

Wash Your Hands! Wash Your Hands! Wash Your Hands!

Use the following hand-washing recipe:

1. If the paper-towel dispenser requires you to touch the handle or lever, the first step should be to crank down the paper towel. Let the paper towel hang there. Do not do this if the paper towel touches and cross-contaminates with the wall or the waste container.

2. Wet your hands (100°F/37.8°C).

3. Add soap.

4. Scrub for 20 seconds.

   ★ Don't forget your nails, thumbs, and between your fingers!

   ★ Some regulators require nailbrushes.

5. Rinse.

6. Dry with a paper towel.

   ★ **Put on gloves** if touching ready-to-eat food.

   ★ If exiting a restroom, wash your hands again when you reenter the kitchen to avoid any contamination that may have occurred when you exited the bathroom and touched the handle to the door.

When Do You Wash Your Hands and Change Your Gloves?

- ★ After going to the bathroom
- ★ Before and after food preparation
- ★ After touching your hair, face, or any other body parts
- ★ After scratching your scalp
- ★ After rubbing your ear
- ★ After touching a pimple
- ★ After wiping your nose and using a tissue
- ★ After sneezing and coughing into your hand
- ★ After drinking, eating, or smoking
- ★ After touching your apron or uniform

- ★ After touching the telephone or door handle
- ★ After touching raw food and before touching ready-to-eat products
- ★ After cleaning and handling all chemicals
- ★ After taking out the trash
- ★ After touching any non-food-contact surfaces
- ★ Every 4 hours during constant use
- ★ After touching a pen
- ★ After handling money
- ★ After receiving deliveries
- ★ Before starting your shift

■ NO BARE-HAND CONTACT

You must not touch **ready-to-eat (RTE)** foods with bare hands. RTE foods are foods that are exactly that: "ready to eat," like bread, pickles, lunch meats, and cheese. These foods can be handled with gloves, deli paper, tongs, and utensils. One in three people do not wash their hands after using the bathroom, but gloves help stop cross-contamination via the fecal/oral route. This is why there is no bare-hand contact with RTE food.

However, some foodservice operators strongly disagree with the no-bare-hand contact of ready-to-eat food. In food safety training programs, trainers frequently hear the reasons why operators have to touch RTE with their bare hands:

- ★ "I've been in business for decades and I have never gotten anyone sick."
- ★ "It takes too much time if you have to wash your hands between every customer. People will be waiting forever!"
- ★ "If you are the only one making the food and using the cash register, it's not realistic."
- ★ "I am being forced to spend my hard-earned money on smallwares and gloves. Do you know the cost of gloves, serving utensils, spoons and tongs these days?"
- ★ "The gloves make the employees wash their hands less."
- ★ "The gloves are hard to get on after you wash your hands."
- ★ "The gloves don't fit right and they make my hands hot."
- ★ "I don't want to change."

A regulatory agency may consider your request and allow bare-hand contact with ready-to-eat food under the requirements of Form 1-D (application for barehand contact procedures). As the person in charge, the choice is yours: smallwares, gloves, additional training, a minor inconvenience versus possible contamination with fecal matter. The choice is yours, knowing that one out of three people do not wash their hands after using the restroom. Food safety is about leadership, being a role model, and making sound decisions. Following is FDA Form 1-D (www.cfsan.fda.gov/~acrobat/fc05-a7.pdf) to present to your health inspector in lieu of washing your hands. Good luck!

FORM 1-D	Application for Bare Hand Contact Procedure (As specified in Food Code ¶ 3-301.11(D))

Please type or print legibly using black or blue ink

1. **Establishment Name:** _____

2. **Establishment Address:** _____

3. **Responsible Person:** _____ **Phone:** _____
                                  Legal Representative   Business

4. **List Procedure and Specific Ready-To-Eat-Foods** to be considered for use of bare hand contact with ready-to-eat foods:

   _____

   _____

   _____

5. **Handwashing Facilities**:

   (a) There is a handwashing sink located immediately adjacent to the posted bare hand contact procedure and the hand sink is maintained in accordance with provisions of the Code. (§ 5-205.11, § 6-301.11, § 6-301.12, § 6-301.14)   ☐ YES ☐ NO   (Include diagram, photo or other information)

   (b) All toilet rooms have one or more handwashing sinks in, or immediately adjacent to them, and the sinks are equipped and maintained in accordance with provisions of the Code. (§ 5-205.11, § 6-301.11, § 6-301.12, § 6-301.14)   ☐ YES ☐ NO

6. **Employee Health Policy:** The written employee health policy must be attached to this form along with documentation that food employees and conditional employees acknowledge their responsibilities. (§ 2-201.11, § 2-201.12, § 2-201.13)

7. **Employee Training**: Provide documentation that food employees have received training in:

   - The risks of contacting the specific ready-to-eat foods with bare hands
   - Personal health and activities as they relate to diseases that are transmissible through food.
   - Proper handwashing procedures to include how, when, where to wash, & fingernail maintenance. (§ 2-301.12, § 2-301.14, § 2-301.15, § 2-302.11)
   - Prohibition of jewelry. (§ 2-303.11)
   - Good hygienic practices. (§ 2-401.11, § 2-401.12)

8. **Documentation of Handwashing Practices:** Provide documentation that food employees are following proper handwashing procedures prior to food preparation and other procedures as necessary to prevent cross-contamination during all hours of operation when the specific ready-to-eat foods are prepared or touched with bare hands.

9. **Documentation of Additional Control Measures:** Provide documentation to demonstrate that food employees are utilizing two or more of the following control measures when contacting ready-to-eat foods with bare hands:

   - Vaccination against hepatitis A for food employees including initial booster shots or documented medical evidence that a food employee has had a previous illness from hepatitis A virus;
   - Double handwashing;
   - Use of nailbrushes;
   - Use of hand antiseptic after handwashing;
   - Incentive programs such as paid leave encouraging food employees not to work when they are ill; or
   - Other control measures approved by the regulatory authority.

<u>Statement of Compliance</u>:

I certify all of the following: All food employees are individually trained in the risks of contacting ready-to-eat foods with bare hands, personal health and activities as they relate to diseases that are transmissible through food, proper handwashing procedures, prohibition of jewelry, and good hygienic practices. A record of this training is kept on site. I understand that bare hand contact with ready-to-eat food is prohibited except for those items listed in section four (4) above. A handwashing sink is located immediately adjacent to the posted bare hand contact procedure. All handwashing sinks are maintained with hot water, soap, and drying devices. I understand that documentation is needed for handwashing practices and additional control measures. I understand that records to document handwashing are kept current and kept on site.

SIGNATURE:_____   DATE _____
               (Signature of legal representative of the facility listed above)

Regulatory Authority (RA) Use Only:

Permit Number: _____

File Review Conducted on History of Handwashing Compliance: ☐ Yes ☐ No

Site Visit Conducted ☐ Yes ☐ No   Comments: _____

☐ Approved: Effective Date: _____   RA name _____

☐ Not Approved: Reason for Denial: _____

_____

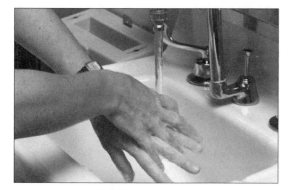

Courtesy PhotoDisc/Getty Images.

★ STAR KNOWLEDGE EXERCISE: WASHING HANDS SOP

In the space below, provide the information related to the instructions, monitoring, corrective action, verification, and record keeping needed for the SOP for washing hands.

SOP: Washing Hands (Exercise)

Purpose: To prevent foodborne illness by contaminated hands.

Scope: This procedure applies to foodservice employees who handle, prepare, or serve food.

Key Words: Hand washing, cross-contamination

Instructions:

1.

2.

3.

4.

5.

6.

7.

8.

9.

10.

Monitoring:

Corrective Action:

Verification and Record Keeping:

Date Implemented:	_____	By:	_____
Date Reviewed:	_____	By:	_____
Date Revised:	_____	By:	_____

■ DO NOT CROSS-CONTAMINATE

Between Raw and RTE or Cooked Foods

Raw food is food that needs to be cooked before it is eaten, like raw meat and eggs. As mentioned previously, **ready-to-eat food** is food that doesn't need to be cooked and is ready to be eaten, like a sandwich roll or lettuce. **Cooked food** is food that has been properly cooked by reaching a specific temperature for an appropriate amount of time, like a cooked hamburger. Once food has been properly cooked, it is now considered a ready-to-eat food.

Food-Contact Surfaces

Cross-contamination occurs when raw food touches or shares contact with ready-to-eat and/or cooked foods. If you touch the walk-in (refrigerator/cooler) door handle, or a pen, or the telephone, and then make a sandwich, this is cross-contamination. Cross-contamination is using the same knife to cut both chicken and rolls. If raw chicken is stored in the refrigerator above lettuce and the chicken juice drips onto the lettuce, this is cross-contamination.

To avoid cross-contamination:

★ Properly store raw food below ready-to-eat food (chicken below lettuce).
★ Never mix food products when restocking.
★ Properly clean and sanitize utensils, equipment, and surfaces.
★ Clean and sanitize work areas when changing from raw food preparation to RTE food preparation.
★ Never store any food near any chemicals.

Between Tasks

It is critical to change gloves, wash hands, and use clean and sanitized utensils, cutting boards, and work surfaces between tasks to prevent contamination. Here are some ways to place a barrier between you and the cross-contamination.

One system to help prevent cross-contamination is to use color-coding. For instance, you can use different-colored gloves for different jobs. This system makes it easy to differentiate food-handling jobs from non-food-handling jobs. Ask your manager if your company has a SOP for gloves. Here are some examples of color-coding gloves:

★ Use clear gloves for food preparation.
★ Use blue gloves for fish.
★ Use yellow gloves for poultry.
★ Use red gloves for beef.
★ Use purple gloves for cleaning and for non-food-contact surfaces.

A similar practice designates different cloths and containers and color-codes them to separate food and non-food-contact surfaces. For example:

★ Use a white cloth for food-contact surfaces.
★ Use a blue cloth for non-food-contact surfaces.
★ Use a green container for cleaning (water and soap).
★ Use a red container for sanitizing (water and sanitizer).
★ You could also color-code cutting boards, knives, containers, and gloves.

STAR KNOWLEDGE EXERCISE: USING SUITABLE UTENSILS WHEN HANDLING RTE FOODS SOP

In the space provided, include information related to instructions, monitoring, corrective action, verification, and record keeping needed for using suitable utensils when handling ready-to-eat foods.

SOP: Using Suitable Utensils When Handling RTE Foods (Exercise)

Purpose: To prevent foodborne illness due to hand-to-food cross-contamination.

Scope: This procedure applies to foodservice employees who prepare, handle, or serve food.

Key Words: Ready-to-eat food, cross-contamination

Instructions:

1.

2.

3.

4.

5.

6.

7.

8.

9.

10.

Monitoring:

Corrective Action:

Verification and Record Keeping:

Date Implemented: _____	By: _____
Date Reviewed: _____	By: _____
Date Revised: _____	By: _____

Dress Code Violations

Be a leader. Be a role model. Follow these rules to avoid violating the dress code:

★ Cover all cuts and burns with a bandage *and* a glove if you have an injury on your hand.

★ Wear your hat or proper hair restraint.

★ Wear clean, closed-toe shoes with rubber soles.

★ Take a bath or shower every day.

★ Always have clean and neat hair.

★ Properly groom fingernails and hands.

★ Do not wear nail polish or false nails.

★ Do not wear rings, necklaces, watches, bracelets, dangly or hoop earrings, or facial piercings. According to the 2005 FDA Model Food Code the **only exceptions** are that a plain wedding band may be worn, a medical alert necklace can be tucked under the shirt, or a medical alert ankle bracelet can be used.

★ Do not chew gum.

★ Only eat, drink, and smoke in designated areas.

Courtesy PhotoDisc/Getty Images.

★ Do not touch your hair, your face, or any body parts when handling or serving food.

★ Remove aprons before leaving the food preparation areas.

★ Wear a clean apron and uniform at all times.

★ Never take your apron into the bathroom.

Sampling Food

Cross-contamination can occur when you are sampling food at your workstation. You should never eat at your workstation unless you are taste-testing the food that you are preparing. Here are some proper ways to sample food:

★ Use a single-use spoon. **Do not double dip!** *Single-use* means exactly that—only one taste per single-use spoon. **OR** Take a small dish and ladle a small portion of the food into the small dish. Put down the ladle. Step away from the pan or pot. Taste the food. Then place items in the dirty dish area of your establishment.

★ Always wash your hands and return to work.

Courtesy Corbis Digital Stock.

STAR KNOWLEDGE EXERCISE: PERSONAL HYGIENE SOP

In the space below, provide information for instructions, monitoring, corrective action, verification, and record keeping needed for a personal hygiene SOP.

SOP: Personal Hygiene (Exercise)

Purpose: To prevent contamination of food by foodservice employees.

Scope: This procedure applies to foodservice employees who handle, prepare, or serve food.

Key Words: Personal hygiene, cross-contamination, contamination

Instructions:

1.

2.

3.

4.

5.

6.

7.

8.

9.

10.

Monitoring:

Corrective Action:

Verification and Record Keeping:

Date Implemented: _____ By: _____

Date Reviewed: _____ By: _____

Date Revised: _____ By: _____

■ POTENTIALLY HAZARDOUS FOODS: TIME/TEMPERATURE CONTROL FOR SAFETY OF FOOD (PHF/TCS)

Potentially Hazardous Foods

A **potentially hazardous food (PHF)** is any food capable of allowing germs to grow rapidly. PHFs have the potential to cause foodborne illness outbreaks. They are usually **moist** (like watermelon), have lots of **protein** (like dairy and meat), and don't have very high or very low acidity (**neutral acidity**). Adding lemon juice or vinegar to foods slows the growth of the germs.

Potentially hazardous food requires strict time and temperature controls to stay safe. Food has been **time/temperature abused** anytime it has been in the **temperature danger zone (TDZ) (41°F to 135°F or 5°C to 57.2°C)** for too long. (More on the TDZ in the next section.) Potentially hazardous foods must be checked often to make sure that they stay safe. The caution sign includes a clock and thermometer to stress the importance of monitoring time **and** temperature. The clock is the reminder to check food at regular time intervals (such as every 2 or 4 hours). The thermometer required must be properly calibrated, cleaned, and sanitized. These are general guidelines describing the qualities of potentially hazardous foods. Scientists have developed much more specific criteria for identifying potentially hazardous foods.

Here is a list of potentially hazardous foods (PHFs):

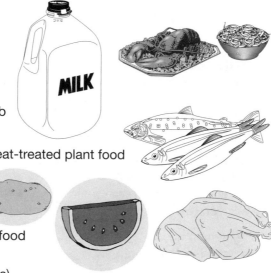

- ★ Milk and milk products
- ★ Shell eggs
- ★ Fish
- ★ Poultry
- ★ Shellfish and crustaceans
- ★ Meats: beef, pork, and lamb
- ★ Baked or boiled potatoes
- ★ Cooked rice, beans, and heat-treated plant food (cooked vegetables)
- ★ Garlic-and-oil mixtures
- ★ Sprouts/sprout seeds
- ★ Tofu and other soy-protein food
- ★ Synthetic ingredients (i.e., soy in meat alternatives)
- ★ Sliced melons

Time/Temperature Control for Safety of Food (TCS)

TCS is the new scientific criteria for foodservice operators whose regulators adopt the 2005 FDA Model Food Code. With these guidelines there is *no easy list of foods* to use as PHF because regulations are based on the characteristics of food. Characteristics involve the interaction of such factors as the following:

- ★ Salt concentration
- ★ Preservatives
- ★ Free available chlorine
- ★ Viscosity
- ★ Humidity
- ★ Oxygen
- ★ pH
- ★ Titratable acidity (**Titratable acidity** measures all the various acids present. An example is measuring the quantity of alkali needed to neutralize the components of a given amount of milk products and milk, and is expressed as percentage of lactic acid. This test is used to evaluate milk quality and to examine the progress of fermentation in cheese and fermented milks.)
- ★ Moisture
- ★ Water activity (a_w)
- ★ Time
- ★ Temperature

Some of these qualities are easy to measure; others will require new tools or gauges to measure. The change in measuring factors and the interaction between factors will help us make sure we are correctly identifying potentially hazardous foods so we can properly control them to be safe. As scientists get more precise in refining (or expanding) the criteria for PHFs, operators need to adapt at the same time. This is one of the reasons the 2005 FDA Model Food Code is a changing, dynamic, adaptive guide that is updated every 4 years.

Anyone whose HACCP plan and its success is based on the 2005 FDA Model Food Code will be impacted and will use these latest guidelines. This includes schools, processors, packagers, and foodservice establishments. The goal of this change is to provide foodservice operators with additional tools to provide useful information to confirm if the food needs time/temperature control for safety (TCS).

Although the criteria to identify foods that need TCS (also known as PHF/TCS) has expanded, it may actually decrease the amount of foods that actually need time/temperature control for safety. For example, the previous definition of PHFs listed only moisture, protein, and neutral acidity as the combination of ingredients to classify a food as PHF. But now, using a matrix and tables of refined values, and the interaction of water activity and pH values and processing methods or preservatives used, we can more correctly identify food that need TCS.

This implementation of new scientific data may seem like more work, but the result is safer food, even giving us better quality of food. Decades ago, in order to kill the trichinosis parasites often found in pork, it was common practice to overcook pork to such a degree that when we ate it, it was dry, hard, and flavorless. Our advances in raising domestic pigs and scientific research now allows us to cook pork to 145°F (62.8°C) for 15 seconds to ensure a juicy pork chop that is still safe to eat. Is checking the temperature with a calibrated thermometer any more difficult than cutting the meat open to see if the pork is completely cooked? Of course not. This is the advantage that science-based HACCP gives us.

The matrix in Tables A and B provide a scientific approach for the interaction of water activity and pH values. These interaction tables were included with the definition of foods that require time/temperature control for safety to limit pathogen growth or toxin formation. The matrix is easy to use with the supporting tables in the appendix of this book. Simply reference the water activity and the pH of the food product in the tables, then compare it to Tables A and B. If the food is not in the table, ask your manufacturer to assist you with the water activity and pH levels in the foods you use.

In foodservice operations, sometimes an operator adds ice to quickly cool food. Sometimes water is added to reconstitute the food or to improve the quality. Remember, each time you alter the product, this impacts the characteristics of the food.

TCS Matrix

Table A. Interaction of pH and a_w for control for spores in food heat-treated to destroy vegetative cells and subsequently packaged

a_w values	pH values		
	4.6 or less	**> 4.6–5.6**	**> 5.6**
≤0.92	non-PHF*/non-TCS food**	non-PHF/non-TCS food	non-PHF/non-TCS food
> 0.92–.95	non-PHF/non-TCS food	non-PHF/non-TCS food	PA***
> 0.95	non-PHF/non-TCS food	PA	PA

* PHF means Potentially Hazardous Food
** TCS food means Time/Temperature Control for Safety food
*** PA menas Product Assessment required

Table B. Interaction of pH and a_w for control of vegetative cells and spores in food not heat-treated or heat-treated but not packaged

a_w values	pH values			
	< 4.2	**4.2–4.6**	**< 4.6–5.0**	**> 5.0**
0.88	non-PHF*/ non-TCS food**	non-PHF/ non-TCS food	non-PHF/ non-TCS food	non-PHF/ non-TCS food
0.88–0.90	non-PHF/ non-TCS food	non-PHF/ non-TCS food	non-PHF/ non-TCS food	PA***
> 0.90–0.92	non-PHF/ non-TCS food	non-PHF/ non-TCS food	PA	PA
> 0.92	non-PHF/ non-TCS food	PA	PA	PA

* PHF means Potentially Hazardous Food
** TCS food means Time/Temperature Control for Safety food
*** PA menas Product Assessment required

Water Activity of Foods

Water activity (a_w) is a critical factor that determines shelf life. While temperature, pH, and several other factors can influence if and how fast organisms will grow in a product, water activity may be the most important factor in controlling spoilage. Water activity refers to the availability of water in a food or beverage and thus the amount of water that is available to microorganisms. Pure water has an activity level of 1.00. Crackers have a water activity of 0.10. Most bacteria, for example, do not grow at water activities below 0.91, and most molds cease to grow at water activities below 0.80. By measuring water activity, you can predict which microorganisms will and will not be potential sources of spoilage. Various food products are highlighted in the table in the appendix.

pH of Foods

The pH and/or acidity of a food are generally used to determine processing requirements and for regulatory purposes. **pH** is the symbol for a measure of the degree of acidity or alkalinity of a solution based on a scale of 0 to 14. Values between 0 and 7 indicate acidity, and values between 7 and 14 indicate alkalinity. The value for pure distilled water is 7, which is considered neutral. Electrodes and pH meters are available from various manufacturers. To assist readers in determining the pH levels of different products, the appendix lists the approximate ranges of pH values.

The minimum pH for Campylobacter spp. to grow is 4.9, the optimum condition is between 6.5 and 7.5, and the maximum pH for growth is 9.0. Eighty percent of all poultry have Campylobacter spp., and poultry has an approximate water activity of 0.99 to 1.00 and a pH range of 6.2 to 6.4. But you must factor the water activity ranges that can support Campylobacter spp. The minimum is 98, and the optimum is 99. So, the higher the water activity, the better the conditions for bacterial growth. This is also the reason why poultry is a potentially hazardous food where time/temperature control is required for the safety of food.

In summary, it is important to understand and manage water activity and pH in controlling the growth of known pathogens. The tables in the appendix are provided for a simple and clearer understanding of pathogens that threaten the foods you prepare and serve. Identifying potentially hazardous foods/temperature control for safe foods helps you avoid putting your operation at risk.

Reduced Oxygen Packaging (ROP) Foods

In the 2005 FDA Model Food Code, reduced oxygen packaging (ROP) encompasses a large variety of packaging methods where the internal environment of the package contains a controlled oxygen level (typically 21 percent at sea level), including vacuum packaging (VP), modified atmosphere packaging (MAP), controlled atmosphere packaging (CAP), cook chill processing (CC), and sous vide (SV, French for "under vacuum"). Using ROP methods in food establishments has the advantage of providing extended shelf life to many foods because it inhibits spoilage organisms that are typically aerobic.

Most foodborne pathogens are anaerobic or facultative anaerobes able to multiply under either aerobic or anaerobic conditions. Therefore, special controls are necessary to control their growth. Refrigerated storage temperatures of 41°F (5°C) may be adequate to prevent growth and/or toxin production of some pathogenic microorganisms, but Clostridium botulinum and Listeria monocytogenes are able to multiply well below 41°F (5°C). For this reason, Clostridium botulinum and Listeria monocytogenes become the pathogens of concern for ROP. Controlling their growth will control the growth of other foodborne pathogens as well.

Some ROP refrigerated foods eliminate some of the preparation steps of foods usually prepared in a foodservice operation. These food items are packaged to extend shelf life. While the packaging inhibits growth of spoilage organisms, it may promote growth of pathogenic bacteria such as Clostridium botulinum and Listeria monocytogenes if the foods are time/temperature abused or served beyond the recommended "use-by" dates. Receiving and storage temperatures are critical for these products.

Examples of these foods are sous vide and MAP foods, as explained in the following:

Types of ROP

★ **Vacuum packaging (VP).** The process in which air is removed from a package of food and the package is hermetically sealed so that a vacuum remains inside the package. Consider lettuce: When stored in the refrigerator, it has a normal shelf life of 3 to 6 days; however, with VP, the shelf life is 2 weeks.

★ **Modified atmosphere packaged (MAP) foods.** Food is partially processed or lightly cooked before being put into a pouch or other container and sealed. Depending on the type of food product, the MAP process uses special gases or mixtures of gases with different properties. MAP specifically includes the reduction in the proportion of oxygen, total replacement of oxygen, or an increase in the proportion of other gases such as carbon dioxide or nitrogen. MAP is used to extend the shelf life and maintain quality food products. Cooked poultry from a manufacturer without MAP has a 5- to 15-day shelf life; however, with the use of MAP, the shelf life is extended to 21 to 30 days. These foods should be received and stored at temperatures of 41°F (5°C) or below.

★ **Controlled atmosphere packaging (CAP) foods.** CAP packaging modifies food so that until the package is opened, its composition is different from air, and continuous control of that atmosphere is maintained, such as by using oxygen scavengers (chemicals placed directly into the packaging wall that absorb oxygen that permeates into the package over time) or a combination of the total replacement of oxygen-respiring foods (i.e., meat and seafood), and impermeable packaging material. The food product is packaged in a laminate or film, and then the atmosphere inside the pack is controlled. After the pack is sealed or vacuum packed, the laminate or film prevents further transmission of gases in or out of the food package, extending the shelf life and ensuring a quality product. Food products where CAP is used include dry fruits, yeast, spices, cereals, rice, fish, meat, and cheese.

★ **Cook chill processing (CC).** Cook chill packaging occurs when cooked food is hot-filled into impermeable bags that have the air expelled and are then sealed or crimped closed. The bagged food is rapidly chilled and refrigerated at temperatures that inhibit the growth of pathogens. CC processing is commonly used for soups, sauces, and meats.

★ **Sous vide (SV).** SV packaged food is put into a package raw or partially cooked and then hermetically sealed in an impermeable bag, cooked in the bag, rapidly chilled, and refrigerated at temperatures that inhibit the growth of pathogens. These foods should be received and stored at temperatures of 41°F (5°C) or below. Foodservice operators are practicing cooking food such as short ribs and beef cheeks at low temperatures in vacuum-packed plastic bags.

ROP packaging has both concerns and benefits. The concerns include the following:

★ Refrigeration may be the only barrier to pathogenic growth.

★ Bacteria such as Clostridium botulinum and Listeria monocytogenes may not be killed.

★ Competing spoilage organisms may be killed. Spoilage organisms inhibit the growth of pathogenic bacteria by competing with them. You can see evidence of spoilage organisms, whereas the presence of pathogens (disease-causing organisms) is not noticeable.

★ Pathogens and their spores may not be destroyed.

★ An anaerobic condition is created, favoring the growth of pathogens such as Clostridium botulinum and Listeria monocytogenes.

★ Cooking may make food more favorable to pathogen growth.

Some benefits of ROP include the following:

★ Higher quality of food

★ Consistent food products

★ Labor savings

★ Menu flexibility

★ Convenience

★ Extended shelf life

★ Food safety

Besides scientific analysis, there are still standard operating procedures that must be followed with other foods. For example, washing fruits and vegetables is critical for their safety. Here is an exercise for standard operating procedures for washing fruits and vegetables.

STAR KNOWLEDGE EXERCISE: WASHING FRUITS AND VEGETABLES SOP

In the space below, list the directions for instructions, monitoring, corrective action, verification, and record keeping needed for washing fruits and vegetables.

SOP: Washing Fruits and Vegetables (Exercise)

Purpose: To prevent or reduce risk of foodborne illness or injury by contaminated fruits and vegetables.

Scope: This procedure applies to foodservice employees who handle, prepare, or serve food.

Key Words: Fruits, vegetables, cross-contamination, washing

Instructions:

1.

2.

3.

4.

5.

6.

7.

8.

9.

10.

Monitoring:

Corrective Action:

Verification and Record Keeping:

Date Implemented: _____ By: _____

Date Reviewed: _____ By: _____

Date Revised: _____ By: _____

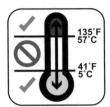

◼ TEMPERATURE DANGER ZONE (TDZ)

Be Safe—Monitor Time and Temperature!

This symbol means no food should stay between **41°F and 135°F (5°C to 57.2°C)**, as this is the **temperature danger zone (TDZ)**. Germs and bacteria grow and multiply very, very fast in this zone. If a PHF/TCS stays in the temperature danger zone of 41°F to 135°F (or 5°C to 57.2°C) for more than 4 hours, it is **time/ temperature abused** and can make people very sick. That is why cold food must be kept cold at 41°F (5°C) or lower and hot food must be kept hot at 135°F (57.2°C) or above. It is important to practice **temperature control (TC)** to make sure foods are not time/temperature abused.

Since foods should not sit on the counter in the TDZ for more than 4 accumulated hours, you should put food away as soon as possible. In the 2005 FDA Model Food Code, there is a new exception to the 4-hour rule. If the internal temperature of food is 41°F (5°C) or lower, once it is removed from TC cold holding, it can remain out of TC for up to **6 hours** as long as the internal product temp does not go above 70°F (21.1°C).

Check holding units (ovens/refrigerators/freezers/warmers/serving lines) at regular intervals to ensure food safety. For example, if the steam table has been accidentally unplugged, it could result in the food temperature dropping to 120°F (48.9°C). If the last time you took the temperature of the food on the table was less than 4 hours ago, you can reheat the food to 165°F (73.9°C) for 15 seconds within 2 hours and continue to serve the product. But if the last time you took the temperature was more than 4 hours ago, then you MUST discard all the foods that are time/temperature abused. This unsafe food can make anyone who eats it sick.

Here is something to think about. What is the temperature of a healthy human?

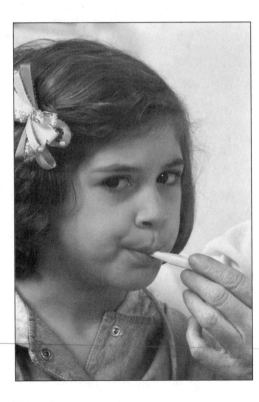

Courtesy PhotoDisc/Getty Images.

If you answered 98.6°F (37°C), you are correct. But note that 98.6°F (37°C) is right in the middle of the temperature danger zone. Our bodies are ideal for germs because we are in the TDZ! Germs love people! Those germs will be transferred to people's food if you are not careful. That is why controlling time and temperature and maintaining good personal hygiene are keys to the success of food safety.

★ STAR KNOWLEDGE EXERCISE: TIME-/DATE-MARKING FOOD SOP

In the space below, provide information for instructions, monitoring, corrective action, verification, and record keeping needed for date-marking food.

SOP: Time-/Date-Marking Food (Exercise)

Purpose: To ensure appropriate rotation of ready-to-eat food to prevent or reduce foodborne illness from Listeria monocytogenes.

Scope: This procedure applies to foodservice employees who handle, prepare, store, or serve food.

Key Words: Ready-to-eat food, potentially hazardous food, date marking, cross-contamination

Instructions:

1.

2.

3.

4.

5.

6.

7.

8.

9.

10.

Monitoring:

Corrective Action:

Verification and Record Keeping:

Date Implemented: _____ By: _____

Date Reviewed: _____ By: _____

Date Revised: _____ By: _____

Checking Food Temperatures with Calibrated Thermometers

What is the point of checking temperatures if you have no clue whether the thermometer is working properly? Calibrated thermometers ensure temperatures of food are correct. There are many types of thermometers. Here are some types common to foodservice:

★ Bimetallic

★ Thermistor (digital)

★ Thermocouple

★ Disposable temperature indicators (t-stick)

★ Infrared with Probe

Thermometers must be checked during **every shift** for correct calibration. The simple act of either dropping a thermometer on the floor or banging the thermometer against a prep table can knock the thermometer out of calibration. All food must be checked with a properly calibrated thermometer. Follow these steps to calibrate a bimetallic stemmed thermometer, the most commonly used thermometer in the foodservice industry.

ICE-POINT METHOD

Step 1

Fill the container with crushed ice and water

Step 2

Submerge sensing area of stem or probe for 30 seconds

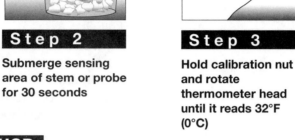

Step 3

Hold calibration nut and rotate thermometer head until it reads 32°F (0°C)

BOILING-POINT METHOD

Step 1

Bring a deep pan of water to a boil

Step 2

Submerge sensing area of stem or probe for 30 seconds

Step 3

Hold calibration nut and rotate thermometer head until it reads 212°F (100°C)

Thermometer Tips:
Store several clean thermometers in a convenient location in a container filled with sanitizer solution. The sanitizer solution should be checked every 4 hours to verify concentration.

Properly Thaw Foods

Often we need to thaw food prior to starting the cooking process. How many times have you thought, "We can pull the turkeys from the freezer and let them sit on the worktable to thaw?" Sitting frozen food on the counter to thaw is **not** a safe food-handling practice. Food needs to safely move through the TDZ as it thaws. There are four safe methods for thawing food:

Method 1:   **Thaw in the refrigerator.** As foods thaw, they may produce extra liquid. Be sure to place PHFs on the lowest shelves of the refrigerator giving consideration to the minimum internal cooking temperatue. Always store in a pan or on a tray to avoid cross-contamination.

Method 2:   **Thaw in running water.** Foods to be thawed under running water must be placed in a sink with running water at 70°F (21.1°C) or cooler. The sink must be open to allow the water to push the microorganisms off the food and flow down the drain. Do not allow the sink to fill with water.

Method 3:   **Cooking.** Frozen food can be thawed by following the cooking directions for the product. Frozen food may take longer to cook depending on the size and type of product.

Method 4:   **Microwave.** Food can be thawed using the microwave if it will then be immediately cooked. When thawing food in the microwave, remember that there will be uneven thawing and some of the food may have started to cook, taking some of the food into the TDZ. This is why you must finish the cooking process immediately after microwave-thawing.

■ COOK ALL FOODS THOROUGHLY

Each PHF/TCS has a minimum internal cooking temperature that must be reached and held for 15 seconds to ensure that it is safe and does not make anyone sick.

Minimum Internal Cooking Temperatures

Here are some minimum internal cooking temperatures to keep in mind when preparing food:

165°F (73.9°C) for 15 Seconds for:

★ Leftover foods and all reheated foods.

★ Poultry and wild game.

★ Stuffed products, including pasta.

★ Combining already cooked and raw PHF products (casseroles).

★ Foods cooked in a microwave; then let sit for 2 minutes.

155°F (68.3°C) for 15 Seconds for:

★ Ground product: fish, beef, and pork.

★ Flavor-injected meats.

★ Eggs for hot holding and later service (buffet service).

145°F (62.8°C) for 15 Seconds for:

★ Fish and shellfish.

★ Chops/steaks of veal, beef, pork, and lamb.

★ Fresh eggs and egg products for immediate service.

★ Roasts to 145°F (62.8°C) for **4 minutes**. (**Note:** The internal temperature needs to be held for longer than 15 seconds because it is a large product that is thick and dense.) For alternate roast temperatures, see tables below and on page 56.

135°F (57.2°C) or 15 Seconds for:

★ RTE foods.

★ Commercially processed products.

★ Vegetables that are to be held hot.

★ Hot holding for all PHF, cooked vegetables, and fruits.

Alternate roast temperatures from the 2005 FDA Model Food Code

Time/Temperature Ranges for Roast Chart I		
Cook to selected internal temperature and hold for specific time (seconds) to destroy organisms.		
°F	**°C**	**Time in Seconds**
158°F	70°C	0 seconds
157°F	69.4°C	14 seconds
155°F	68.3°C	22 seconds
153°F	67.2°C	34 seconds
151°F	66.1°C	54 seconds
149°F	65°C	85 seconds
147°F	63.9°C	134 seconds
Note: Alternate roast temperatures from the 2005 FDA Model Food Code.		

Time/Temperature Ranges for Roast Chart II

Cook to selected internal temperature and hold for specific time (minutes) to destroy organisms.

°F	°C	Time in Minutes
145°F	62.8°C	4 minutes
144°F	62.2°C	5 minutes
142°F	61.1°C	8 minutes
140°F	60°C	12 minutes
138°F	58.9°C	18 minutes
136°F	57.8°C	28 minutes
135°F	57.2°C	36 minutes
133°F	56.1°C	56 minutes
131°F	55°C	89 minutes
130°F	54.4°C	112 minutes

Note: Alternate roast temperatures from the 2005 PDA Model Food Code.

STAR KNOWLEDGE EXERCISE: COOKING SOP

In the space below, provide information for the instructions, monitoring, corrective action, verification, and record keeping needed for cooking foods.

SOP: Cooking (Exercise)

Purpose: To prevent foodborne illness by ensuring that all foods are cooked to the appropriate internal temperature.

Scope: This procedure applies to foodservice employees who prepare or serve food.

Key Words: Cross-contamination, temperatures, cooking

Instructions:

1.

2.

3.

4.

5.

6.

7.

8.

9.

10.

Monitoring:

Corrective Action:

Verification and Record Keeping:

Date Implemented: _____ By: _____

Date Reviewed: _____ By: _____

Date Revised: _____ By: _____

■ COLD HOLDING

Be Safe—Monitor Time and Temperature!

Here are time and temperature food safety rules:

★ In **cold-holding/self-service bars (refrigeration)**, store all cold food below 41°F (5°C).

★ Check temperatures of the food in cold holding a minimum of every 4 hours with a calibrated, clean, and sanitized thermometer.

★ Always keep food out of the TDZ.

■ HOT HOLDING

Be Safe—Monitor Time and Temperature!

Here are time and temperature food safety rules:

★ In **hot-holding/self-service bars (steam table)**, store all hot food above 135°F (57.2°C).

★ Check temperatures of the food in hot holding a minimum of every 4 hours with a calibrated, clean, and sanitized thermometer.

★ Always keep food out of the TDZ.

★ STAR KNOWLEDGE EXERCISE: HOLDING HOT AND COLD PHF/TCS FOODS SOP

In the space below, provide information for the instructions, monitoring, corrective action, verification, and record keeping needed for holding hot and cold foods.

SOP: Holding Hot and Cold PHF/TCS Foods (Exercise)

Purpose: To prevent foodborne illness by holding foods at the correct temperature.

Scope: This procedure applies to foodservice employees who handle, prepare, or serve food.

Key Words: Cross-contamination, hot holding, cold holding, storage, temperature

Instructions:

1.

2.

3.

4.

5.

6.

7.

8.

9.

10.

Monitoring:

Corrective Action:

Verification and Record Keeping:

Date Implemented:	_____ By: _____
Date Reviewed:	_____ By: _____
Date Revised:	_____ By: _____

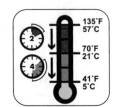

■ COOLING FOOD

Two-stage cooling allows potentially hazardous food to be in the temperature danger zone for more than 4 hours only if these strict guidelines are followed. Cool hot food from **135°F to 70°F (57.2°C to 21.1°C)** within 2 hours; you then have an additional 4 hours to go from **70°F to 41°F (21.1°C to 5°C)** or lower for a maximum total cool time of 6 hours. Note: If the food does not reach 70°F (21.1°C) within 2 hours, you must immediately reheat to 165°F (73.9°C) and then begin the cooling process again from that point.

Cool food as quickly as possible. Keep in mind that 6 hours is the maximum amount of time only if you reach **70°F (21.1°C) within 2 hours**. The reason you can have an additional 4 hours is that the food moves through the most dangerous section of the TDZ within the first 2 hours. **Less time is better.** Your goal when cooling food is to move food as quickly as possible through the TDZ. Proper ways to cool food quickly are as follows:

★ Use a clean and sanitized ice paddle.

★ Stir food to release the heat.

★ Use an ice bath.

★ Add ice as an ingredient.

★ Use a quick-chill unit such as a blast chiller.

★ Separate food into smaller portions or thinner pieces.

Once food has cooled to 70°F (21.1°C), it should be placed in the refrigerator as follows:

★ Place food in shallow stainless steel pans (no more than 4 inches deep).

★ Make sure pan cover is loose to allow the heat to escape.

★ Place pans on top shelves in refrigeration units.

★ Position pans so air circulates around them. *(Be cautious not to overload refrigerator tray racks.)*

★ Monitor food to ensure two-stage cooling process. Cool hot food from 135°F to 70°F (57.2°C to 21.1°C) within 2 hours or less; you then have up to an additional 4 hours to go from 70°F to 41°F (21.1°C to 5°C) or lower for a maximum total cool time of 6 hours.

The refrigerator and freezer are not designed to cool hot food. The warmest temperature food to be placed in a refrigerator is 70°F (21.1°C); in a freezer, 41°F (5°C). If hot food is placed in a refrigerator unit, the refrigerator unit will work harder and warm the other foods that are supposed to be cold, ruining both the foods and the expensive refrigeration unit.

STAR KNOWLEDGE EXERCISE: COOLING FOOD SOP

In the space below, provide information for the instructions, monitoring, corrective action, verification, and record keeping needed for holding hot and cold PHF/TCS.

SOP: Cooling Food (Exercise)

Purpose: To prevent foodborne illness by properly cooling food.

Scope: This procedure applies to foodservice employees who handle, prepare, or serve food.

Key Words: Cooling, temperature

Instructions:

1.

2.

3.

4.

5.

6.

7.

8.

9.

10.

Monitoring:

Corrective Action:

Verification and Record Keeping:

Date Implemented: _____ By: _____

Date Reviewed: _____ By: _____

Date Revised: _____ By: _____

Reheating

The goal of reheating is to move food as quickly as possible through the TDZ. It is critical when reheating food to cook the food to a minimum of 165°F (73.9°C) for 15 seconds within 2 hours. If food takes longer than 2 hours to reheat, it must be discarded or thrown away. Use steam when possible to reheat food and not dry heat. **Never use hot-holding equipment to reheat food**, because this equipment is not designed for that purpose. Hot-holding equipment is designed to hold the temperature once the food is hot.

STAR KNOWLEDGE EXERCISE: REHEATING FOOD SOP

In the space below, provide the information for the instructions, monitoring, corrective action, verification, and record keeping needed for reheating.

SOP: Reheating Food (Exercise)

Purpose: To prevent foodborne illness by reheating food properly.

Scope: This procedure applies to foodservice employees who handle, prepare, or serve food.

Key Words: Reheating, temperature

Instructions:

1.

2.

3.

4.

5.

6.

7.

8.

9.

10.

Monitoring:

Corrective Action:

Verification and Record Keeping:

Date Implemented: _____ By: _____

Date Reviewed: _____ By: _____

Date Revised: _____ By: _____

■ WASH, RINSE, SANITIZE

Clean and Sanitize! "Sparkle!"

Follow proper cleaning and sanitizing food safety rules, as outlined in the following.

What Is Cleaning?

Cleaning is removing the dirt you can see on a surface. The expectation is for everything to "sparkle!" A sparkling-clean foodservice operation impresses each customer. Clean all surfaces, equipment, and utensils every 4 hours or when they become soiled or when they no longer sparkle. You can use detergents or solvents or scraping to clean.

What Does It Mean to Sanitize?

Sanitizing is reducing the unseen germs on a surface to a **safe** level.

1. Sanitize all things that come in contact with food, including utensils, cutting boards, and prep tables.
2. Clean and sanitize at minimum every **4 hours**.

3. You can sanitize with water that is at least 180°F/82.2°C (dishwashing machines) or use a chemical sanitizer.

You must follow the proper SOP for your foodservice operation. There are five important points to remember:

★ Always use a **sanitizer test strip** when preparing a sanitizer solution.

★ Use **separate cloths** for food surfaces like a prep table and non-food surfaces like a wall or floor.

★ Use a **designated sink system** like the three-compartment sink to clean and sanitize dishes and utensils. Never clean and sanitize dishes in the hand-washing or food preparation sinks. Mop water can only be emptied into the utility sink or the toilet, never in the three-compartment sink.

★ Always **keep chemicals and food products separate**. Never receive cleaning products and chemicals on the same pallet with food. Remember, to prevent cross-contamination, you should always be alert when receiving your deliveries. This is a serious concern associated with the potential chemical contamination of deliveries.

★ Keep copies of the **Material Safety Data Sheets** (MSDS) for each chemical used on premises.

How Do You Set Up a Three-Compartment Sink?

Step 1:   Clean and sanitize entire sink and drainboards before starting.
Step 2:   Scrape and rinse dirty dishes.
Step 3:   Wash at 110°F (43.3°C) with soapy water.
Step 4:   Rinse at 110°F (43.3°C) with clear water.
Step 5:   Sanitize using your SOP.
Step 6:   Air-dry.

Chemicals

When you are working with chemicals (poisonous and toxic), prerequisite programs need to be in place and followed in order to prevent a chemical contamination or misuse. All operations should have a chemical management plan clearly stating that only approved chemicals necessary to the establishment should actually be in the establishment. Your program needs to outline the specific **storage** procedures of all chemicals in a secure cabinet away from all food and utensils. The prerequisite programs also must illustrate the **proper use** of the chemicals. According to the 2005 FDA Model Food Code, medicines necessary for the health of employees may be allowed in a food establishment, but they should be labeled and stored to prevent contamination of food and food-contact surfaces. All chemicals should bear a legible manufacturer's label. All spray bottles, buckets, and working containers must be **clearly labeled** with the common name of the chemical. Make sure your chemicals are not being shipped on the same truck or pallet as food. Have your supplier verify that food is protected from chemical contamination during shipment. Any food that has been cross-contaminated with chemicals should be rejected or discarded immediately. Discard chemicals used in working containers and mop buckets in an appropriate service sink to prevent contamination of food and food-contact surfaces.

STAR KNOWLEDGE EXERCISE: CLEANING AND SANITIZING SOP

In the space provided, list the instructions, monitoring, corrective action, verification, and record keeping needed for the Cleaning and Sanitizing SOP.

SOP: Cleaning and Sanitizing (Exercise)

Purpose: To prevent foodborne illness by improper cleaning and sanitizing.

Scope: This procedure applies to foodservice employees who handle, prepare, or serve food.

Key Words: Cleaning, sanitizing, cross-contamination

Instructions:

1.

2.

3.

4.

5.

6.

7.

8.

9.

10.

Monitoring:

Corrective Action:

Verification and Record Keeping:

Date Implemented: _____ By: _____

Date Reviewed: _____ By: _____

Date Revised: _____ By: _____

★ PEST CONTROL

The pest control prerequisite program should define the established pest control system and the use of a pest control log, address the grounds surrounding your facility, block pest access into the facility, and monitor and maintain facilities on a regular basis. This prerequisite program needs to work in cooperation with your licensed pest control operator (PCO). Your PCO should thoroughly survey and inspect the interior and exterior of your facility, develop a customized program based on detected problems, execute an effective treatment plan, and, finally, put preventative measures in place to maintain control over pests that may enter your facility during deliveries. Let's go into further detail for your pest control prerequisite program:

★ **Establish a pest control system.** This system must include routine inspections as well as daily cleaning as described in the master cleaning schedule. The chemicals and pesticides used to control the pests must be locked in a cabinet. Keep a copy of the corresponding Material Safety Data Sheets (MSDS) on the premises. Dispose of empty containers according to local regulations and manufacturer's directions.

★ **Use a pest control log.** Document the pest control actions taken by your PCO. File records with HACCP records.

★ **Maintain the grounds in good condition.** Properly landscape the outside of your facility. For example, tall grass provides excellent nesting and hiding places for pests.

★ **Block access of pests into facility.** Keep windows and doors closed. Seal all openings into the facility to prevent future entry of pests, rodents, or pets. Ensure that any rodents/pests that may have entered the facility are no longer present. Remove dead pests and sanitize any food-contact surfaces that have come in contact with pests.

★ **Monitor and maintain facilities regularly.** Condition of the physical structure of the establishment should be in compliance with local building and occupancy codes in a manner that does not compromise the safe and sanitary handling of food and equipment and the safety of employees. Conduct routine inspection of the facility.

SERVING FOOD AND OPERATING SELF-SERVICE BARS

SERVING FOOD

Can you answer **YES** to any of these questions?

★ Are dinner plates and/or coffee cups stacked when serving food and drink to customers?

★ Are the server's fingers on the edge of the plate in the food?

★ Are customers served, tables cleared, the phone answered, and payments taken without washing hands?

★ Are rolls, unwrapped butter, and uneaten garnishes (pickles) from plates recycled?

★ Are utensils, towels, or order pads stored in pockets or waistband?

If you answered YES to any of these questions, now is the time to start serving food safely. Don't let food safety end in the kitchen! Everyone in the foodservice operation can play an important role in food safety. Servers should never stack dinner plates and cups on top of one another, or on arms, or carry too many in one hand. It is a surefire way to cross-contaminate foods. Today's customer is more educated and will be more aware of servers who bus tables and then touch plates or glasses as they deliver food without washing their hands between each task. This same customer is also aware of the server who answers the phone, writes down an order, prepares the food, rings the register, and collects the money while wearing the same pair of gloves he or she wore when going through the same routine for the three previous customers. Think about the serving practices that need improvement in your foodservice operation.

It is very important not to **reuse** food like rolls, unwrapped butter, and uneaten pickle garnishes. The safest rule to follow is that any food that leaves your foodservice or your control should **never** be served to another customer.

You should always carry all utensils by the handle, carry all glasses by the side, and carry all plates from the bottom. Do not store utensils and cloths in your pockets or in the waistband of your clothes.

STAR KNOWLEDGE EXERCISE: SERVING FOOD SOP

In the space below, provide information for the instructions, monitoring, corrective action, verification, and record keeping needed for a Service SOP.

SOP: Serving Food (Exercise)

Purpose: To prevent foodborne illness by serving food properly.

Scope: This procedure applies to foodservice employees who handle, prepare, or serve food.

Key Words: Hand washing, cross-contamination

Instructions:

1.

2.

3.

4.

5.

6.

7.

8.

9.

10.

Monitoring:

Corrective Action:

Verification and Record Keeping:

Date Implemented: _____ By: _____

Date Reviewed: _____ By: _____

Date Revised: _____ By: _____

■ SELF-SERVICE AREAS

Although many customers enjoy the convenience of self-service areas, some of them are not aware of the dangers that they inflict on those salad bars, beverage stations, condiment areas. Since we can not "train" a customer on how to properly and safely use these areas, signage is important. "Please use a new plate when returning to the salad bar."

It is also very important that employees are vigilant in maintaining all self-service areas.

★ Ensure every item on a buffet has its own serving utensil.

★ Every item must be labeled, so customers won't "taste it" for identification purposes.

★ Maintaining the cleanliness of the area without contaminating any food with cleaner.

★ Monitoring the customers at the self-service area and eliminating any item that has been touched, tasted, or returned because it has now been contaminated.

STAR KNOWLEDGE EXERCISE: SELF-SERVICE AREAS SOP

In the space provided, list the directions for instructions, monitoring, corrective action, verification, and record keeping needed for self-service areas.

SOP: Self-Service Areas (Exercise)

Purpose: To prevent foodborne illness by preventing contamination in self-service areas.

Scope: This procedure applies to foodservice employees who handle, prepare, or serve food.

Key Words: Hand washing, cross-contamination

Instructions:

1.

2.

3.

4.

5.

6.

7.

8.

9.

10.

Monitoring:

Corrective Action:

Verification and Record Keeping:

Date Implemented: _____ By: _____

Date Reviewed: _____ By: _____

Date Revised: _____ By: _____

PREREQUISITE PROGRAMS STAR CONCLUSION

If you know how to handle the following situations, you will ensure that your food is safe. As mentioned previously, foods may become unsafe accidentally because of cross-contamination, poor personal hygiene, improper cleaning and sanitizing, and time/temperature abuse. It's important that you keep the food, yourself, other employees, and your customers safe at all times. Now that you've read this chapter, let's see how much you know about food safety.

★ ARE YOU A FOOD SAFETY "SUPERSTAR"?

Match the following scenarios to the area(s) of concern related to the food safety practices listed.

A. Prevent cross-contamination.

B. Demonstrate proper personal hygiene.

C. Use proper cleaning and sanitizing procedures.

D. Monitor and take corrective action for time and temperature.

Situation/scenario	Identify the area(s) of concern (A-D) from the list above	What do you do to correct the situation? Make it a "REAL" solution!
1. A serving utensil falls on the floor.		
2. An employee wore a dirty uniform to work.		
3. You are stocking shelves and notice the date on a carton of shell eggs is expired.		
4. It is 3 p.m. and you find a pan of sausage on the prep table left out since breakfast. Breakfast ended at 11 a.m.		

(continues)

Situation/scenario	Identify the area(s) of concern (A-D) from the list on page 73	What do you do to correct the situation? Make it a "REAL" solution!
5. A fellow manager comes out of the bathroom. You see her tying her apron.		
6. A customer returns a meatball sandwich because it is cold.		
7. The sanitizer solution is supposed to be 200 ppm (parts per million). You see a new coworker set up the three-compartment sink, but he uses too much sanitizer.		
8. A customer is allergic to fish. The server tells the cook that her customer has a fish allergy. The customer ordered a hamburger but there is only one spatula on the production line that is used for everything.		
9. A employee is angry with a disgruntled customer and spits on the customer's plate.		
10. Right before closing, you have your employee cleaning the walls in the food service area. A customer rushes in and places an order.		

SUMMARY OF FOOD SAFETY STANDARD OPERATING PROCEDURES (SOPs)

The following USDA SOPs were completed during the various exercises in this Star Point. These completed versions can be compared to what you included. You can customize completed SOP examples as needed.

- ★ Purchasing (p. 75)
- ★ Receiving (p. 78)
- ★ Storing Food Properly (p. 79; also called Storage)
- ★ Washing Hands (p. 81)
- ★ Personal Hygiene (p. 82)
- ★ Using Suitable Utensils When Handling Foods (p. 84)
- ★ Washing Fruits and Vegetables (p. 85)
- ★ Time/Date-Marking Food (p. 86)
- ★ Cooking PHF/TCS Foods (p. 87)
- ★ Holding Hot and Cold PHF/TCS Foods (p. 89)
- ★ Cooling (p. 91)
- ★ Reheating (p. 92)
- ★ Cleaning and Sanitizing (p. 93)
- ★ Serving Food (p. 94)
- ★ Self-Service Areas (p. 95)

SOP: Purchasing (Sample)

Purpose: To prevent contamination of food and to ensure safe foods are served to customers by purchasing food products from approved suppliers. These suppliers must be approved by appropriate regulatory services.

Scope: This procedure applies to foodservice managers who purchase foods from approved suppliers.

Key Words: Approved suppliers, regulatory services

Instructions:

Contact regulatory services to ensure you are purchasing foods from approved suppliers. To find out if a supplier is approved, call

- ★ CDC Food Safety Office—404-639-2213 or visit www.cdc.gov
- ★ EPA—202-272-0167 or visit www.epa.gov
- ★ FSIS—888-674-6854 or visit www.fsis.usda.gov
- ★ FDA—888-463-6332 or visit www.cfsan.fda.gov

1.	Domestic/imported food (including produce, bottled water, and other foods) *but not meat and poultry*	★ Evidence of regulatory oversight: copy of suppliers, local enforcement agency permit, state or federal registration or license, or a copy of the last inspection report.
		★ Third-party audit results [many vendors now provide third-party guarantees, including NSF International or American Institute of Baking (AIB)].
		★ Microbiological or chemical analysis/testing results.
		★ Person-in-the-plant verification (i.e., chain food facilities may have their own inspector monitor food they buy).
		★ Self-certification (guarantee) by a wholesale processor based on HACCP.
		★ For raw agricultural commodities such as produce, certification of Good Agricultural Practices or membership in a trade association such as the United Fresh Fruit and Vegetable Association.
		★ A copy of a wholesale distributor or processor's agreement with its suppliers of food safety compliance.
2.	Domestic/imported meat, poultry, and related products such as meat- or poultry-containing stews, frozen foods, and pizzas	★ USDA mark on meat or poultry products
		★ Registration of importers with USDA
3.	Fish and Fish Products	★ Evidence of regulatory oversight: copy of suppliers' local enforcement agency permit, state or federal registration or license, or a copy of the last inspection report
		★ Third-party audit results
		★ Person-in-the-plant verification
		★ Self-certification (guarantee) by a wholesale processor based on HACCP
		★ A copy of a wholesale distributor or processor's agreement with its suppliers of HACCP compliance.
		★ U.S. Department of Commerce (USDC) approved list of fish establishments and products located at seafood.nmfs.noaa.gov
4.	Shellfish	★ Shellfish tags
		★ Listing in current *Interstate Certified Shellfish Shippers* publication
		★ Gulf oyster treatment process verification if sold between April 1 and October 31 (November 1 to March 31 certification may be used in lieu of warning signs)
		★ USDC-approved list of fish establishments and products located at seafood.nmfs.noaa.gov

5.	Drinking water (nonbottled water)	★	A recent certified laboratory report demonstrating compliance with drinking water standards
		★	A copy of the latest inspection report
6.	Alcoholic beverages	★	Third-party audit results
		★	Self-certification (guarantee) by a wholesale processor based on HACCP
		★	Person-in-the-plant verification
		★	Evidence of regulatory oversight: copy of suppliers' local enforcement agency permit, state or federal registration or license, or a copy of the last inspection report
		★	A copy of a wholesale distributor or processor's agreement with its suppliers of food safety compliance

Monitoring:

1. Inspect invoices or other documents to determine approval by a regulatory agency.
2. Food service managers should be encouraged to make frequent inspections of the suppliers' on-site facilities, manufacturing facilities, and processing plants/farms. Inspections determine cleanliness standards and ensure that HACCP plans are in place.

Corrective Action:

Food service purchasing managers must find a new supplier if the supplier is not approved by the above regulatory services.

Verification and Record Keeping:

The food service purchasing manager will maintain all documentation from food suppliers. Documentation must be maintained for three years plus the current year.

Date Implemented: _____ By: _____

Date Reviewed: _____ By: _____

Date Revised: _____ By: _____

SOP: Receiving Deliveries (Sample)

Purpose: To ensure that all food is received fresh and safe when it enters the foodservice operation, and to transfer food to proper storage as quickly as possible

Scope: This procedure applies to foodservice employees who handle, prepare, or serve food.

Key Words: Cross-contamination, temperatures, receiving, holding, frozen goods, delivery

Instructions:

1. Train foodservice employees who accept deliveries on proper receiving procedures.
2. Schedule deliveries to arrive at designated times during operational hours.
3. Post the delivery schedule, including the names of vendors, days and times of deliveries, and drivers' names.
4. Establish a rejection policy to ensure accurate, timely, consistent, and effective refusal and return of rejected goods.
5. Organize freezer and refrigeration space, loading docks, and storerooms before receiving deliveries.
6. Before deliveries, gather product specification lists and purchase orders, temperature logs, calibrated thermometers, pens, and flashlights, and be sure to use clean loading carts.
7. Keep receiving area clean and well lighted.
8. Do not touch ready-to-eat foods with bare hands.
9. Determine whether foods will be marked with the date of arrival or the "use-by" date, and mark accordingly upon receipt.
10. Compare delivery invoice against products ordered and products delivered.
11. Transfer foods to their appropriate locations as quickly as possible.

Monitoring:

1. Inspect the delivery truck when it arrives to ensure that it is clean, free of putrid odors, and organized to prevent cross-contamination. Be sure refrigerated foods are delivered on a refrigerated truck.
2. Check the interior temperature of refrigerated trucks.
3. Confirm vendor name, day and time of delivery, as well as driver's identification before accepting delivery. If the driver's name is different than what is indicated on the delivery schedule, contact the vendor immediately.
4. Check frozen foods to ensure that they are all frozen solid and show no signs of thawing and refreezing, such as the presence of large ice crystals or liquids on the bottom of cartons.
5. Check the temperature of refrigerated foods.
   ★ For fresh meat, fish, dairy, and poultry products, insert a clean and sanitized thermometer into the center of the product to ensure a temperature of 41°F (5°C) or below.
   ★ For packaged products, insert a food thermometer between two packages, being careful not to puncture the wrapper. If the temperature exceeds 41°F (5°C), it may be necessary to take the internal temperature before accepting the product.
   ★ For eggs, the interior temperature of the truck should be 45°F (7.2°C) or below.

6. Check dates of milk, eggs, and other perishable goods to ensure safety and quality.

7. Check the integrity of food packaging.

8. Check the cleanliness of crates and other shipping containers before accepting products. Reject foods that are shipped in dirty crates.

Corrective Action:

Reject the following:

★ Frozen foods with signs of previous thawing

★ Cans that have signs of deterioration—swollen sides or ends, flawed seals or seams, dents, or rust

★ Punctured packages

★ Expired foods

★ Foods that are out of the safe temperature zone or deemed unacceptable by the established rejection policy

Verification and Record Keeping:

The designated team member needs to record temperatures and corrective actions taken on the delivery invoice or on the receiving log. The foodservice manager will verify that foodservice employees are receiving products using the proper procedure by visually monitoring receiving practices during the shift and reviewing the receiving log at the close of each day. Receiving and corrective action logs are kept on file for a minimum of 1 year.

Date Implemented: _____	By:	_____
Date Reviewed: _____	By:	_____
Date Revised: _____	By:	_____

SOP: Storage (Sample)

Purpose: To ensure that food is stored safely and put away as quickly as possible after it enters the foodservice operation

Scope: This procedure applies to foodservice employees who handle, prepare, or serve food.

Key Words: Cross-contamination, temperatures, storing, dry storage, refrigeration, freezer

Instructions: Answers

1. Freezer temperature is –10°F to 0°F (–23.3°C to –17.8°C).

2. Refrigerator temperatures are between 36°F and 39°F (2.2°C to 3.9°C).

3. Dry storage temperatures are between 50°F and 70°F (10°C and 21.1°C). Humidity is between 50 percent and 60 percent.

4. Record freezer and refrigerator temperatures on the appropriate log at the specific times.

5. Use a calibrated, clean, and sanitized thermometer (+/–2°F) or (+/– 1°C).

6. FIFO (first-in, first-out) procedures are used for storage. All items are dated upon delivery.

7. All food stored in the freezer, refrigerator, and dry storage must be covered, dated, labeled, and stored 6 inches off the floor.

8. Potentially hazardous foods are stored no more than 7 days at 41°F (5°C) from the date of preparation.

9. Cooked and ready-to-eat foods are stored above raw foods. Store other foods based on minimum internal cooking temperature.

10. Always store food in its original container as long as it is clean, dry, and intact. If not, notify your manager, director, or person in charge.

11. Never put food in an empty chemical container. Never put chemicals in an empty food container.

12. Pesticides and chemicals are stored in a secure area away from food handling and storage areas. Never store pesticide and chemicals in the food preparation and storage areas. Always store them in a locked cabinet.

Monitoring:

1. Check frozen foods to ensure that they are all frozen solid and show no signs of thawing and refreezing, such as the presence of large ice crystals or liquids on the bottom of cartons.

2. Check the temperature of refrigerated foods.

   a. For fresh meat, fish, dairy, and poultry products, insert a clean, sanitized, and calibrated thermometer into the center of the product to ensure a temperature of 41°F (5°C) or below.

   b. For packaged products, insert a food thermometer between two packages, being careful not to puncture the wrapper. If the temperature exceeds 41°F (5°C), it may be necessary to take the internal temperature.

   c. For eggs, the ambient temperature should be 45°F (7.2°C) or below.

3. Check dates of milk, eggs, and other perishable goods to ensure safety and quality.

4. Check the integrity of food packaging.

5. Check the cleanliness of the dry storage room, refrigeration units, and freezer units.

Corrective Action:

1. Discard the following:

   a. Frozen foods with signs of previous thawing

   b. Cans that have signs of deterioration—swollen sides or ends, flawed seals or seams, dents, or rust

   c. Punctured packages

   d. Expired foods

   e. Foods that are out of the safe temperature zone or deemed unacceptable by the established rejection policy

Verification and Record Keeping:

Record temperature and corrective action on the appropriate storage log or chart. The foodservice manager will verify that foodservice employees are storing products using the proper procedure by visually monitoring storing practices during the shift and reviewing the storage log at the close of each day. Storage and corrective action logs are kept on file for a minimum of 1 year.

Date Implemented: _____ By: _____

Date Reviewed: _____ By: _____

Date Revised: _____ By: _____

SOP: Washing Hands (Sample)

Purpose: To prevent foodborne illness by contaminated hands

Scope: This procedure applies to personnel who handle, prepare, and serve food.

Keywords: Hand washing, cross-contamination

Instructions:

1. Train any individual that prepares or serves food on proper hand washing. Training may include showing a hand-washing video and demonstrating proper hand-washing procedure.

2. Post hand-washing signs or posters in a language understood by all foodservice staff near all hand-washing sinks, in food preparation areas, and in restrooms.

3. Use designated hand-washing sinks for hand washing only. Do not use food preparation, utility, and dish-washing sinks for hand washing. Do not use hand-washing sinks for food preparation, utility, or dish washing.

4. Provide warm running water, soap, and a means to dry hands. Provide a waste container at each hand-washing sink or near the door in restrooms.

5. Make hand-washing sinks accessible in any area where employees are working.

6. Hands must be washed:
   - ★ Before starting work
   - ★ During food preparation
   - ★ When moving from one food preparation area to another
   - ★ Before putting on or changing gloves
   - ★ After using the toilet
   - ★ After sneezing, coughing, or using a handkerchief or tissue
   - ★ After touching hair, face, or body
   - ★ After smoking, eating, drinking, or chewing gum or tobacco
   - ★ After handling raw meats, poultry, or fish
   - ★ After any cleanup activity such as sweeping, mopping, or wiping counters
   - ★ After touching dirty dishes, equipment, or utensils
   - ★ After handling trash
   - ★ After handling money
   - ★ After any time the hands may become contaminated

7. Use paper towel to open door when exiting the restroom.

8. Follow proper hand-washing procedures as indicated below:
   - ★ Wet hands and forearms with warm, running water (at least 100°F / 37.8°C) and apply soap.
   - ★ Scrub lathered hands and forearms, under fingernails, and between fingers for at least 10 to 15 seconds. Rinse thoroughly under warm running water for 5 to 10 seconds.
   - ★ Dry hands and forearms thoroughly with single-use paper towels.
   - ★ Dry hands for at least 30 seconds if using a warm-air hand dryer.
   - ★ Turn off water by using paper towels.

9. Follow FDA recommendations when using hand antiseptics. These recommendations are as follows:

   ★ Use hand antiseptics only after hands have been properly washed and dried.

   ★ Use only hand antiseptics that comply with the 2005 FDA Model Food Code. Confirm with the manufacturers that the hand antiseptics used meet these requirements. Use hand antiseptics in the manner specified by the manufacturer.

Monitoring:

A designated employee will visually observe the hand-washing practices of the foodservice staff during all hours of operation. In addition, the designated employee will visually observe that hand-washing sinks are properly supplied during all hours of operation.

Corrective Action:

Employees that are observed not washing their hands at the appropriate times or using the proper procedure will be asked to wash their hands immediately. Employee will be retrained to ensure proper hand-washing procedure.

Verification and Record Keeping:

The foodservice manager will complete the Food Safety Checklist daily to indicate that monitoring is being conducted as specified. Maintain Food Safety Checklist for a minimum of 1 year.

Date Implemented: _____	By: _____
Date Reviewed: _____	By: _____
Date Revised: _____	By: _____

SOP: Personal Hygiene (Sample)

Purpose: To prevent contamination of food by foodservice employees

Scope: This procedure applies to foodservice employees who handle, prepare, or serve food.

Key Words: Personal hygiene, cross-contamination, contamination

Instructions:

1. Train foodservice employees on the employee health policy (develop SOP for implementing an employee health policy) and on practicing good personal hygiene.
2. Follow the employee health policy.
3. Report to work in good health, clean, and dressed in clean attire.
4. Change apron when it becomes soiled.
5. Wash hands properly, frequently, and at the appropriate times.
6. Keep fingernails trimmed, filed, and maintained so that the edges are cleanable and not rough.

7. Avoid wearing artificial fingernails and fingernail polish.

8. Wear single-use gloves if artificial fingernails or fingernail polish are worn.

9. Do not wear any jewelry except for a plain ring such as a wedding band.

10. Treat and bandage wounds and sores immediately. When hands are bandaged, single-use gloves must be worn.

11. Cover a lesion containing pus with a bandage. If the lesion is on a finger, hand, or wrist, cover with a bandage and finger cot or a bandage and a single-use glove.

12. Eat, drink, use tobacco, or chew gum only in designated break areas where food or food-contact surfaces may not become contaminated.

13. Taste food the correct way:
    a. Place a small amount of food into a separate container.
    b. Step away from exposed food and food-contact surfaces.
    c. Use a teaspoon to taste the food. Remove the used teaspoon and container to the dish room. Never reuse a spoon that has already been used for tasting.
    d. Wash hands immediately.

14. Wear suitable and effective hair restraints while in the kitchen.

15. Follow state and local public health requirements.

Monitoring:

A designated foodservice employee will inspect employees when they report to work to be sure that each employee is following this SOP. The designated foodservice employee will ensure that all foodservice employees are adhering to the personal hygiene policy during all hours of operation.

Corrective Action:

Any foodservice employee found not following this procedure will be retrained at the time of the incident. Affected food will be discarded.

Verification and Record Keeping:

The foodservice manager will verify that foodservice employees are following this policy by visually observing the employees during all hours of operation. The foodservice manager will complete the Food Safety Checklist daily. Foodservice employees will record any discarded food on the Damaged or Discarded Product Log, which will be kept on file for a minimum of 1 year.

Date Implemented: _____	By: _____
Date Reviewed: _____	By: _____
Date Revised: _____	By: _____

SOP: Using Suitable Utensils When Handling Foods (Sample)

Purpose: To prevent foodborne illness due to hand-to-food cross-contamination

Scope: This procedure applies to foodservice employees who prepare, handle, or serve food.

Key Words: Ready-to-eat food, cross-contamination

Instructions:

1. Use proper hand-washing procedures to wash hands and exposed arms prior to preparing or handling food or at any time when the hands may have become contaminated.

2. Do not use bare hands to handle ready-to-eat foods at any time unless washing fruits and vegetables.

3. Use suitable utensils when working with ready-to-eat food. Suitable utensils may include the following:

    a. Single-use gloves

    b. Deli tissue

    c. Foil wrap

    d. Tongs, spoodles, spoons, and spatulas

4. Wash hands and change gloves:

    ★ Before beginning food preparation

    ★ Before beginning a new task

    ★ After touching equipment (such as refrigerator doors) or utensils that have not been cleaned and sanitized

    ★ After contacting chemicals

    ★ When interruptions in food preparation occur, such as when answering the telephone or checking in a delivery

    ★ Handling money

    ★ Anytime a glove is torn, damaged, or soiled

    ★ Anytime contamination of a glove might have occurred

5. Follow state and local public health requirements.

Monitoring:

A designated foodservice employee will visually observe that gloves or suitable utensils are used and changed at the appropriate times during all hours of operation.

Corrective Action:

Employees observed touching ready-to-eat food with bare hands will be retrained at the time of the incident. Ready-to-eat food touched with bare hands will be discarded.

Verification and Record Keeping:

The foodservice manager will verify that foodservice workers are using suitable utensils by visually monitoring foodservice employees during all hours of operation. The foodservice manager will complete the Food Safety Checklist daily. The designated foodservice employee responsible for monitoring will record any discarded food on the Damaged and Discarded Product Log. This log will be maintained for a minimum of 1 year.

Date Implemented: _____ By: _____

Date Reviewed: _____ By: _____

Date Revised: _____ By: _____

SOP: Washing Fruits and Vegetables (Sample)

Purpose: To prevent or reduce risk of foodborne illness or injury by contaminated fruits and vegetables

Scope: This procedure applies to foodservice employees who prepare or serve food.

Key Words: Fruits, vegetables, cross-contamination, washing

Instructions:

1. Train foodservice employees who prepare or serve food on how to properly wash and store fresh fruits and vegetables.
2. Wash hands using the proper procedure.
3. Wash, rinse, sanitize, and air-dry all food-contact surfaces, equipment, and utensils that will be in contact with produce, such as cutting boards, knives, and sinks.
4. Follow manufacturer's instructions for proper use of chemicals.
5. Wash all raw fruits and vegetables thoroughly before combining with other ingredients, including:
   a. Unpeeled fresh fruit and vegetables that are served whole or cut into pieces
   b. Fruits and vegetables that are peeled and cut to use in cooking or served ready-to-eat
6. Wash fresh produce vigorously under cold running water or by using chemicals that comply with the 2005 FDA Model Food Code. Packaged fruits and vegetables labeled as being previously washed and ready-to-eat are not required to be washed.
7. Scrub the surface of firm fruits or vegetables such as apples or potatoes using a clean and sanitized brush designated for this purpose.
8. Remove any damaged or bruised areas.
9. Label, date, and refrigerate fresh-cut items.
10. Serve cut melons within 7 days if held at 41°F (5°C) or below (see SOP for Date-Marking Ready-to-Eat, Potentially Hazardous Food).
11. Do not serve raw seed sprouts to highly susceptible populations such as preschool-age children.
12. Follow state and local public health requirements.

Monitoring:

The foodservice manager will visually monitor that fruits and vegetables are being properly washed, labeled, and dated during all hours of operation. In addition, foodservice employees will check daily the quality of fruits and vegetables in cold storage.

Corrective Action:

Unwashed fruits and vegetables will be removed from service and washed immediately before being served. Unlabeled fresh-cut items will be labeled and dated. Discard cut melons held at 41°F (5°C) or below after 7 days.

Verification and Record Keeping:

The foodservice manager will complete the Food Safety Checklist daily to indicate that monitoring is being conducted as specified in this procedure.

Date Implemented: _____ By: _____

Date Reviewed: _____ By: _____

Date Revised: _____ By: _____

SOP: Time-/Date-Marking Food (Sample)

Purpose: To ensure appropriate rotation of ready-to-eat food to prevent or reduce foodborne illness such as Listeria monocytogenes

Scope: This procedure applies to foodservice employees who prepare, store, or serve food.

Key Words: Ready-to-eat food, potentially hazardous food, date marking, cross-contamination

Instructions:

1. Establish a date-marking system and train employees accordingly. The best practice for a date-marking system would be to include a label with the product name, the day or date, and time it is prepared or opened. Examples of how to indicate when the food is prepared or opened include

    a. Labeling food with a calendar date, such as "cut cantaloupe, 5/26, 8:00 a.m."

    b. Identifying the day of the week, such as "cut cantaloupe, Monday, 8:00 a.m.," or

    c. Using color-coded marks or tags, such as "cut cantaloupe, blue dot, 8:00 a.m." means "cut on Monday at 8:00 a.m."

2. Label ready-to-eat, potentially hazardous foods that are prepared on-site and held for more than 24 hours.

3. Label any processed, ready-to-eat, potentially hazardous foods when opened, if they are to be held for more than 24 hours.

4. Refrigerate all ready-to-eat, potentially hazardous foods at 41°F (5°C) or below.

5. Serve or discard refrigerated, ready-to-eat, potentially hazardous foods within 7 days.

6. Indicate with a separate label the date prepared, the date frozen, and the date thawed of any refrigerated, ready-to-eat, potentially hazardous foods.

7. Calculate the 7-day time period by counting only the days that the food is under refrigeration. For example:

   ★ On Monday, 8/1, lasagna is cooked, properly cooled, and refrigerated with a label that reads, "Lasagna–Cooked–8/1."

   ★ On Tuesday, 8/2, the lasagna is frozen with a second label that reads, "Frozen–8/2." Two labels now appear on the lasagna. Since the lasagna was held under refrigeration from Monday, 8/1 to Tuesday, 8/2, only 1 day is counted toward the 7-day time period.

   ★ On Tuesday, 8/16, the lasagna is pulled out of the freezer. A third label is placed on the lasagna that reads, "Thawed–8/16." All three labels now appear on the lasagna. The lasagna must be served or discarded within 6 days.

8. Follow state and local public health requirements.

Monitoring:

A designated employee will check refrigerators daily to verify that foods are date marked and that foods exceeding the 7-day time period are not being used or stored.

Corrective Action:

Foods that are not date marked or that exceed the 7-day time period will be discarded.

Verification and Record Keeping:

The foodservice manager will complete the Food Safety Checklist daily.

Date Implemented: _____	By:	_____
Date Reviewed: _____	By:	_____
Date Revised: _____	By:	_____

SOP: Cooking PHF/TCS Foods (Sample)

Purpose: To prevent foodborne illness by ensuring that all foods are cooked to the appropriate internal temperature

Scope: This procedure applies to foodservice employees who prepare or serve food.

Key Words: Cross-contamination, temperatures, cooking

Instructions:

1. Train foodservice employees who prepare or serve food on how to use a food thermometer and cook foods using this procedure.

2. If a recipe contains a combination of meat products, cook the product to the highest required temperature.

3. Follow state or local health department requirements regarding internal cooking temperatures.

    If state or local health department requirements are based on the 2005 FDA Model Food Code, cook products to the following temperatures:

    A. 135°F (57.2°C)—Fresh, frozen, or canned fruits and vegetables that are going to be held on a steam table or in a hot box are to be held at 135°F (57.2°C).

    B. 145°F (62.8°C)—Seafood, beef, pork, and eggs cooked to order that are placed onto a plate and immediately served. (Roast 145°F/62.8°C for 4 minutes.)

    C. 155°F (68.3°C) for 15 seconds—Ground products containing beef, pork, or fish, fish nuggets or sticks, eggs held on a steam table, cubed or Salisbury steaks

    D. 165°F (73.9°C) for 15 seconds—Poultry, stuffed fish, pork, or beef, pasta stuffed with eggs, fish, pork, or beef (such as lasagna or manicotti)

Monitoring:

1. Use a clean, sanitized, and calibrated probe thermometer (preferably a thermocouple).
2. Avoid inserting the thermometer into pockets of fat or near bones when taking internal cooking temperatures.
3. Take at least **two** internal temperatures from each batch of food by inserting the thermometer into the thickest part of the product (usually the center).
4. Take at least **two** internal temperatures of each large food item, such as a turkey, to ensure that all parts of the product reach the required cooking temperature.

Corrective Action:

Continue cooking food until the internal temperature reaches the required temperature for the specific amount of time.

Verification and Record Keeping:

Foodservice employees will record product name, time, the two temperatures/times, and any corrective action taken on the Cooking–Reheating Temperature Log.

The foodservice manager will verify that foodservice employees have taken the required cooking temperatures by visually monitoring foodservice employees and preparation procedures during the shift and reviewing, initialing, and dating the temperature log at the close of each day. The Cooking–Reheating Temperature Logs are kept on file for a minimum of 1 year.

Date Implemented: _____	By: _____
Date Reviewed: _____	By: _____
Date Revised: _____	By: _____

SOP: Holding Hot and Cold PHF/TCS Foods (Sample)

Purpose: To prevent foodborne illness by ensuring that all foods are held under the proper temperature

Scope: This procedure applies to foodservice employees who prepare or serve food.

Key Words: Cross-contamination, temperatures, holding, hot holding, cold holding, storage

Instructions:

1. Train foodservice employees who prepare or serve food about proper hot- and cold-holding procedures. Include in the training a discussion of the temperature danger zone.

2. Follow state or local health department requirements regarding required hot- and cold-holding temperatures. If state or local health department requirements are based on the 2005 FDA Model Food Code:
   ★ Hold hot foods at 135°F (57.2°C) or above.
   ★ Hold cold foods at 41°F (5°C) or below.

3. Preheat steam tables and hot boxes.

Monitoring:

1. Use a clean, sanitized, and calibrated probe thermometer to measure the temperature of the food.

2. Take temperatures of foods by inserting the thermometer near the surface of the product, at the thickest part, and at other various locations.

3. Take temperatures of holding units by placing a calibrated thermometer in the coolest part of a hot-holding unit or warmest part of a cold-holding unit.

4. For hot foods held for service:
   a. Verify that the air/water temperature of any unit is at 135°F (57.2°C) or above before use.
   b. Reheat foods in accordance with the Reheating for Hot Holding SOP.
   c. All hot potentially hazardous foods should be 135°F (57.2°C) or above before the food is placed for display or service.
   d. Take the internal temperature of food before placing it on a steam table or in a hot-holding unit and at least every 2 hours thereafter.

5. For cold foods held for service:
   a. Verify that the air/water temperature of any unit is at 41°F (5°C) or below before use.
   b. Chill foods, if applicable, in accordance with the Cooling (PHF/TCS) SOP.
   c. All cold potentially hazardous foods should be 41°F (5°C) or below before placing the food out for display or service.
   d. Take the internal temperature of the food before placing it onto any salad bar, display cooler, or cold serving line and at least every 2 hours thereafter.

6. For cold foods in storage:
   a. Take the internal temperature of the food before placing it into any walk-in cooler or reach-in cold-holding unit.
   b. Chill food in accordance with the Cooling Potentially Hazardous Foods SOP if the food is not 41°F (5°C) or below.
   c. Verify that the air temperature of any cold-holding unit is at 41°F or below before use and at least every 4 hours thereafter during all hours of operation.

Corrective Action:

For hot foods:

1. Reheat the food to 165°F (73.9°C) for 15 seconds if the temperature is found to be below 135°F (57.2°C) and the last temperature measurement was 135°F (57.2°C) or higher and taken within the last 2 hours. Repair or reset holding equipment before returning the food to the unit, if applicable.

2. Discard the food if it cannot be determined how long the food temperature was below 135°F (57.2°C).

For cold foods:

1. Rapidly chill the food using an appropriate cooling method if the temperature is found to be above 41°F (5°C) and the last temperature measurement was 41°F (5°C) or below and taken within the last 2 hours.

2. To rapidly chill the food, place the food in shallow containers (no more than 4 inches deep), cover loosely, and put on the top shelf in the back of the walk-in or reach-in cooler.

    Or use a quick-chill unit like a blast chiller.

    Or stir the food in a container placed in an ice-water bath.

    Or add ice as an ingredient.

    Or separate food into smaller or thinner portions.

    Or use a combination of these methods to cool the food as quickly as possible.

3. Repair or reset holding equipment before returning the food to the unit, if applicable.

4. Discard the food if it cannot be determined how long the food temperature was above 41°F (5°C).

Verification and Record Keeping:

Foodservice employees will record temperatures of food items and document corrective actions taken on the Hot- and Cold-Holding Temperature Log. A designated foodservice employee will record air temperatures of coolers and cold-holding units on the Refrigeration Logs. Foodservice manager will verify that foodservice employees have taken the required holding temperatures by visually monitoring foodservice employees during the shift and reviewing the temperature logs at the close of each day. Maintain the temperature logs for a minimum of 1 year.

Date Implemented: _____ By: _____

Date Reviewed: _____ By: _____

Date Revised: _____ By: _____

SOP: Cooling (Sample)

Purpose: To prevent contamination of food by foodservice employees

Scope: This procedure applies to foodservice employees who handle, prepare, or serve food.

Key Words: Cooling method, quick-chill

Instructions:

1. Potentially hazardous foods must be cooled to 70°F (21.1°C) within 2 hours and to 41°F (5°C) within an additional 4 hours, for a total of 6 hours.

2. Rapidly chill the food using an appropriate cooling method if the temperature is found to be above 41°F (5°C) and the last temperature measurement was 41°F (5°C) or below and taken within the last 2 hours.

3. To rapidly chill the food, place the food in shallow containers (no more than 4 inches deep), cover loosely, and put on the top shelf in the back of the walk-in or reach-in cooler.

    Or use a quick-chill unit like a blast chiller.

    Or stir the food in a container placed in an ice-water bath.

    Or add ice as an ingredient.

    Or separate food into smaller or thinner portions.

    Or use a combination of these methods to cool the food as quickly as possible.

4. Repair or reset holding equipment before returning the food to the unit, if applicable.

5. Discard the food if it cannot be determined how long the food temperature was above 41°F (5°C).

Monitoring:

A designated foodservice employee will inspect employees and storage areas to be sure that each employee is following this SOP. The designated foodservice employee will monitor that all food-service employees are adhering to the storage policy during all hours of operation.

Corrective Action:

Any foodservice employee found not following this procedure will be retrained at the time of the in-cident. Affected food will be discarded.

Verification and Record Keeping:

The foodservice manager will verify that foodservice employees are following this policy by visu-ally observing the employees during all hours of operation. The foodservice manager will complete the Food Safety Checklist daily.

Date Implemented: _____	By:	_____
Date Reviewed: _____	By:	_____
Date Revised: _____	By:	_____

SOP: Reheating (Sample)

Purpose: To prevent contamination of food by foodservice employees.

Scope: This procedure applies to foodservice employees who handle, prepare, or serve food.

Key Words: (PHF/TCS), Ready-to-eat food (RTE), preparation

Instructions:

1. An accurate, calibrated thermometer (+/– 2°F) (+/– 1°C) must be used to take the temperatures of potentially hazardous foods.

2. Discard foods if
   a. They are held in the temperature danger zone (41°F – 135°F or 5°C – 57.2°C) for more than 4 hours.
   b. It cannot be determined how long the food temperature was below 135°F (57.2°C).
   c. They have been cooled too slowly.
   d. They are not reheated to 165°F (73.8°C) (in the thickest part) within 2 hours.

3. Potentially hazardous foods must be cooled to 70°F (21.1°C) within 2 hours and to 41°F (5°C) within an additional 4 hours, for a total of 6 hours.

4. Reheat food to 165°F (73.8°C) or higher for at least 15 seconds within 2 hours if it seems the food will not cool to 70°F (21.1°C) within 2 hours. Serve the food immediately or begin the cooling process; use practical means to speedily cool.

5. If reheated in a microwave, foods must be reheated to an internal temperature of 165°F (73.8°C) and should stand for 2 minutes (this allows the heat to spread evenly throughout the food). Food should be stirred or rotated when possible.

6. Foods are to be labeled with the date and time of preparation before storing.

7. Refrigerated, potentially hazardous, ready-to-eat foods that are held for more than 24 hours after preparation must be used within 7 days or less if the food is held at 41°F (5°C) or lower. However, they must be used within 4 days from the preparation date if they are held at 45°F (7.2°C) or lower.

8. To reduce temperatures, potentially hazardous foods are cooled with an ice paddle, in an ice bath, or in shallow pans.

Monitoring:

A designated foodservice employee will observe employees following this SOP. The designated foodservice employee will monitor that all foodservice employees are adhering to the proper reheating practices during all hours of operation. The Reheating Log will be completed and reviewed.

Corrective Action:

Any foodservice employee found not following this procedure will be retrained at the time of the incident. Affected food will be discarded. The Corrective Action log will be completed.

Verification and Record Keeping:

The foodservice manager will verify that foodservice employees are following this policy by visually observing the employees during all hours of operation. The foodservice manager will complete the Food Safety Checklist daily and review the Reheating and Corrective Action Logs.

Date Implemented: _____ By: _____

Date Reviewed: _____ By: _____

Date Revised: _____ By: _____

SOP: Cleaning and Sanitizing (Sample)

Purpose: To prevent contamination of food by foodservice employees

Scope: This procedure applies to foodservice employees who handle, prepare, or serve food.

Key Words: Kitchenware, fixed equipment, sanitizing, contamination

Instructions:

1. Train foodservice employees on how to:
   a. Properly wash, rinse, and sanitize kitchenware after each use.
   b. Clean equipment that handles potentially hazardous foods at least every 4 hours.
2. Ensure that the third sink of the three-compartment sink is used for sanitizing and that items are sanitized by being immersed in either:
   a. Hot water temperatures that vary based on regulatory requirements from 171°F to 180°F (77.2°C to 82.2°C) for 30 seconds or
   b. A properly mixed chemical sanitizing solution for the recommended time.
3. If using a machine with hot-water sanitizing, the wash water temperatures 150°F to 165°F (65.5°C to 73.9°C) and the sanitizing water 165°F to 194°F (73.9°C to 90°C) temperatures are checked, recorded, and maintained daily. (Temperatures vary depending on type of equipment.)
4. For fixed equipment, removable parts are removed, washed, rinsed, and sanitized by immersion after each use, and nonremovable food-contact surfaces are washed, rinsed, and sanitized with a cloth.
5. Sanitizing solution for worktables is kept in a labeled container and changed at least every 2 hours, frequently depending on use.

Monitoring:

A designated foodservice employee will inspect employees as they work to be sure that each employee is following this SOP. The designated foodservice employee will monitor that all foodservice employees are adhering to the cleaning and sanitizing policy during all hours of operation.

Corrective Action:

Any foodservice employee found not following this procedure will be retrained at the time of the incident. Affected food will be discarded, and affected equipment will be cleaned.

Verification and Record Keeping:

The foodservice manager will verify that foodservice employees are following this policy by visually observing the employees during all hours of operation. The foodservice manager will complete the Food Safety Checklist daily.

Date Implemented: _____ By: _____

Date Reviewed: _____ By: _____

Date Revised: _____ By: _____

SOP: Serving Food (Sample)

Purpose: To prevent contamination of food by foodservice employees who are serving food

Scope: This procedure applies to foodservice employees who handle, prepare, or serve food.

Key Words: Hand washing, cross-contamination, sanitize

Instructions:

1. Hand washing must be monitored and enforced before the food handler is allowed to prepare or serve food.
2. Food handlers must be made aware of poor personal habits, such as touching the mouth, face, hair, dirty apron, or dirty cloth—they may be sources of cross-contamination.

Employees must be trained

1. Not to touch plates, utensils, drinking glasses or cups where the customer's food or mouth will come in contact with the surface.
2. When dishing a customer plate, to wipe drips with a clean cloth or fresh paper towel. The counter cloth should not be used so as to prevent cross-contamination.
3. To use sanitized utensils to handle food, not their bare hands.
4. To use proper serving utensils with long handles so as to prevent handles from coming in contact with food.
5. When scoop and ladles are not in use, to store them in the food with the handle out. Tongs must be stored on a dry, clean surface or in a separate pan.
6. To use gloves for some operations involving handling ready-to-eat foods and for which utensils are not practical, such as sandwich making. Gloves must be changed under the same conditions as hand washing.
7. To be aware of related duties that require hand washing before continuing with service, such as:
   a. Picking up an item from the floor
   b. Handling soiled dishes and linens
   c. Answering the telephone
   d. Handling cash

At self-service stations:

1. Use sneeze guards to protect food.
2. Provide sufficient long-handled utensils so that the handles do not come in contact with the food.
3. Don't overfill containers so that food comes in contact with utensil handles.
4. Require that customers use a fresh plate with each return to the self-service station.
5. Discourage eating or picking with hands at the station.
6. Constantly monitor the customers while they are at the station to prevent cross-contamination of food at the self-service station.

Monitoring:

A designated foodservice employee will inspect employees while they are serving to be sure that each employee is following this SOP. The designated foodservice employee will monitor that all foodservice employees are adhering to the service policy during all hours of operation.

Corrective Action:

Any foodservice employee found not following this procedure will be retrained at the time of the incident. Affected food will be discarded.

Verification and Record Keeping:

The foodservice manager will verify that foodservice employees are following this policy by visually observing the employees during all hours of operation. The foodservice manager will complete the Food Safety Checklist daily.

Date Implemented: _____	By: _____
Date Reviewed: _____	By: _____
Date Revised: _____	By: _____

SOP: Food Safety for Self-Service Areas (Sample)

Purpose: To prevent contamination of food by foodservice employees
Scope: This procedure applies to foodservice employees who handle, prepare, or serve food.
Key Words: PHF/TCS, ice bath, blast chiller

Instructions:

1. Separate raw meat, fish, and poultry from cooked and ready-to-eat food.
2. Monitor customers for unsanitary hygiene practices, such as the following:
   ★ Tasting items.
   ★ Handling multiple breads with their bare hands.
   ★ Putting fingers directly into the food.
   ★ Reusing plates and utensils; instead, hand out fresh plates to customers.

3. Label all food items.

4. Maintain proper temperatures.

   ★ PHF/TCS must be cooled to 70°F (21.1°C) within 2 hours and to 41°F (5°C) within an additional 4 hours, for a total of 6 hours.

   ★ Rapidly chill the food using an appropriate cooling method if the temperature is found to be above 41°F (5°C) and the last temperature measurement was 41°F (5°C) or below and taken within the last 2 hours.

   ★ Place food in shallow containers (no more than 4 inches deep) and *uncovered* on the top shelf in the back of the walk-in or reach-in cooler.

   ★ Use a quick-chill unit such as a blast chiller.

   ★ Stir the food in a container placed in an ice-water bath.

   ★ Add ice as an ingredient.

   ★ Repair or reset holding equipment before returning the food to the unit, if applicable.

   ★ Discard the food if it cannot be determined how long the food temperature was above 41°F (5°C).

   ★ Separate food into smaller or thinner portions.

   ★ When refilling items, never mix old food with new food.

Monitoring:

A designated foodservice employee will inspect employees to be sure that each employee is following this SOP. The designated foodservice employee will monitor all self-service areas during all hours of operation.

Corrective Action:

Any foodservice employee found not following this procedure will be retrained at the time of the incident. Affected food will be discarded.

Verification and Record Keeping:

The foodservice manager will verify that foodservice employees are following this policy by visually observing the employees during all hours of operation. The foodservice manager will complete the Food Safety Checklist daily.

Date Implemented: _____ By: _____

Date Reviewed: _____ By: _____

Date Revised: _____ By: _____:

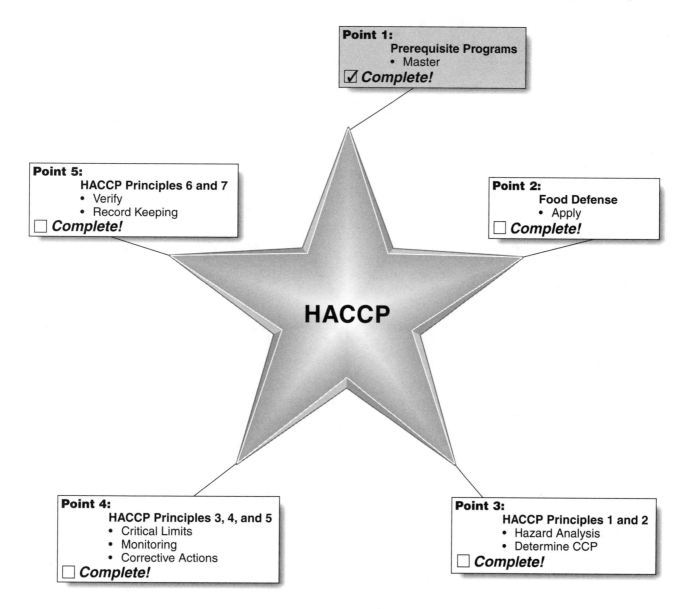

Food Defense

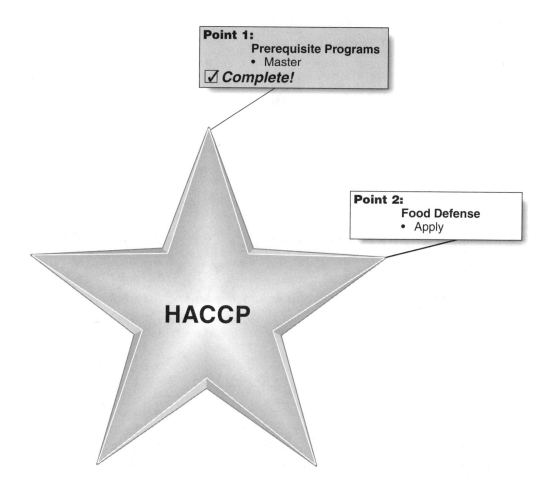

Point 1:
Prerequisite Programs
• Master
☑ *Complete!*

Point 2:
Food Defense
• Apply

HACCP

We in the food industry must concern ourselves with what the U.S. Department of Homeland Security calls **food security** and **food defense standard operating procedures**. As a responsible manager, you should become **more aware** of the safety issues in your operation, the foodservice industry, and the entire country. We must work as a team; each of us must do our part, from the farm to our table, to stop any potential deliberate food contamination. In this chapter, we discuss documented acts of **food terrorism**. It is important for you to recognize that as a food handler, supervisor, manager, director, and leader in the foodservice industry, **it is your responsibility to take action** to prevent acts of food terrorism from occurring at your foodservice establishment.

Star Point Actions: You will learn to

★ Appreciate the federal action taken by the U.S. government to protect our food.

★ Differentiate between food defense, food security, food safety, and alleged contamination or hoaxes.

★ Recognize the importance of food defense in all aspects of the food industry—from the farm to the patron's table.

★ Reverse the "it won't happen to me" attitude.

★ Identify your role in food defense as a responsible leader.

★ Determine how to handle different food defense situations.

★ Identify appropriate food defense standard operating procedures to teach employees when training them in today's world.

★ Investigate food recall procedures and what is needed in your operation.

★ Create a crisis management plan in case a food crisis occurs at your establishment.

FEDERAL ACTION TAKEN TO PROTECT OUR FOOD

The FDA's plan, **"Protecting and Advancing America's Health: A Strategic Action Plan for the 21st Century,"** outlines various steps the agency is taking to respond to the new food defense challenges faced every day in today's world.

■ EVOLUTION OF FOOD DEFENSE

The events of September 11, 2001, reinforced the need to enhance the security of the United States. Congress responded to this security need by passing the **Public Health Security and Bio-terrorism Preparedness and Response Act of 2002** (the Bio-terrorism Act), which President George W. Bush signed into law June 12, 2002. This landmark legislation provided the FDA with new and significant resources to help protect the nation's food supply from the threat of intentional contamination and other food-related emergencies.

The United States Government has taken several important steps toward securing the nation's food supply. On December 17, 2003, President George W. Bush issued the **Homeland Security Presidential Directive (HSPD) 7**. This directive established a national policy for federal departments and agencies to identify and prioritize critical U.S. infrastructures and key resources in an effort to protect the U.S. from terrorist attacks.

The Bush administration determined that due to its vulnerability to a terrorist attack, the food and agriculture sector was a crucial infrastructure to address. As a result, the USDA and the Department of Health and Human Services (HHS) share responsibility for the food and agriculture infrastructure sector. Their responsibilities include the following:

★ Cooperating with key persons

★ Conducting or contributing to vulnerability evaluations

★ Encouraging risk management strategies

★ Identifying, prioritizing, evaluating, remedying, and protecting their respective internal critical infrastructure and key resources

A key specification of **HSPD 7** requires the USDA and HHS to (1) develop indications and early-warning mechanisms for potential food and/or agriculture contamination, (2) integrate their cyber and physical protection plans to protect against food and/or agriculture contamination, and (3) submit regular status reports on private sector coordination to help track the activity of food and agriculture in case contamination has occurred at a given phase from farm to table. Efforts are also currently underway to establish an independent organization that will facilitate better communication of sensitive information between government and industry, since this information may help to protect the food and agriculture sector, and ultimately the consumer.

HSPD 9, issued on January 30, 2004, provides a national policy to defend the U.S. food and agriculture system against terrorist attacks, major disasters, and other emergencies. The USDA, HHS, and the Environmental Protection Agency (EPA) have united to accomplish this by

★ Identifying and prioritizing sector-critical parts of the infrastructure and key resources for achieving protection requirements

★ Developing awareness and early warning systems to recognize vulnerabilities or flaws in the food defense system

★ Decreasing vulnerabilities or flaws at critical production and processing areas

★ Improving screening procedures for domestic and imported products

★ Improving response and recovery procedures if an act of food terrorism has occurred

As a result of this directive, agencies have developed surveillance and monitoring systems for animal disease, plant disease, wildlife disease, as well as food, water, and public health.

■ FOOD DEFENSE VS. FOOD SECURITY VS. FOOD SAFETY VS. HOAXES

Food defense is the protection of food products from **intentional** adulteration/contamination.

Why Is Food Defense Important and Why Should You Take Action?

Terrorist attacks can occur during the various stages of the farm to table chain, which include cultivation of crops, raising livestock, and distributing, processing, retailing, transporting, and storing the food product. It is crucial to practice food defense during all phases of the farm to table chain to ensure that food is safe at all times.

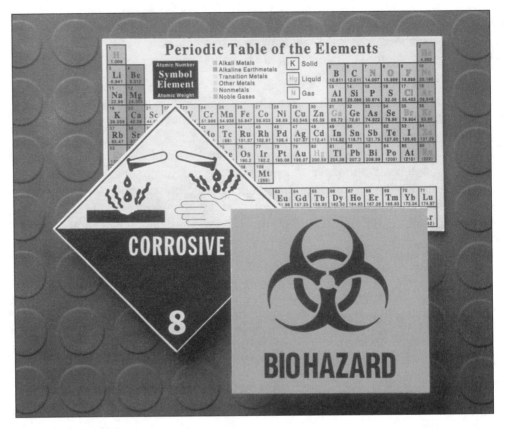

Courtesy PhotoDisc, Inc.

Intentional Food Terrorism Incidents

★ In 2003, a disgruntled supermarket employee contaminated 200 pounds of ground beef with a nicotine-based pesticide. This incident caused a reported 92 people to become ill.

★ In 1993, a disgruntled former employee contaminated a tray of dough-nuts and muffins with Shigella dysenteriae Type 2, which caused 12 em-ployees to suffer severe gastrointestinal illness, and 4 employees to be hospitalized.

★ In 1984, cult members in Oregon added salmonella to several restaurant salad bars in an attempt to affect the outcome of local elections. This incident caused 751 illnesses and 45 people had to be hospitalized.

Food security is best defined by the **World Health Organization** (WHO) as "The implication that all people at all times have both physical and economic access to enough food for an active, healthy life." Internationally, **food security** is defined as a 2-year supply of food for a particular country.

Unlike the rest of the world, the United States originally referred to **food defense** as **food security**. This terminology was shifted only when the U.S. government re-alized that food security was not compatible when the term was used in other parts of the world. While food security and food defense may be used inter-changeably depending on where you are in the world, the common goal is to pre-vent the deliberate contamination of food and to make sure the food we serve is safe.

Unlike food defense, which is the protection of food products from intentional contamination, **food safety** is the protection of food products from **unintentional** contamination.

Unintentional Incidents

- ★ In 2003, hepatitis A caused more than 650 confirmed illnesses and 4 deaths in seven states. All the victims consumed raw or undercooked onions.

- ★ In 1994, Salmonella enteritidis caused 150 confirmed illnesses when the pasteurized ice cream mix produced at one facility became cross-contaminated.

- ★ In 1985, Salmonella typhimurium caused more than 16,000 confirmed illnesses and 17 deaths in six states. All the victims consumed recontaminated pasteurized milk from milk produced at one dairy plant.

◼ HOAXES

Equally damaging are false accusations or fraudulent reports of deliberate contamination of food. Managers must train their employees to be **responsible** employees; if a false alarm is suspected, the person in charge should be notified. An example of a hoax would be if Company XYZ fell victim to a hoax where a customer stated that she found a horrifying object in her food. After further investigation, the horrifying object was actually placed in the food by the customer as a ploy to collect a large sum of money from the company. This physical contaminant hoax is damaging to the employees, the company, the brand, and the foodservice industry. Stories like this negatively affect everyone!

According to the United States General Accounting Office Resources, Community, and Economic Development Division: "Threats of contamination with a biological agent occur infrequently: From October 1995 through March 1999, federal agencies reported receiving three such threats—**two of these were hoaxes**, and the other is still an open investigation." More common are incidents using physical objects to contaminate food.

The **Center for Disease Control and Prevention**, and the National Center for Infectious Diseases, stated that the e-mail rumor circulating about Costa Rican bananas causing the disease necrotizing fasciitis, also known as **flesh-eating bacteria**, is false. Again, hoaxes are costly to the industry and/or establishment related to the allegation, the employees in the industry, and the government agency that investigates the false claim.

◼ WHY IS FOOD DEFENSE IMPORTANT?

Consider the following scenarios that could affect the safety of the food in your establishment:

- ★ You leave the kitchen door open for a long period of time.
- ★ You encounter unusual behavior from a customer or a coworker.
- ★ You do not check to make sure your deliveries are safe and intact.
- ★ There is no designated employee who monitors salad bars and food displays.

★ Chemicals are stored near the food in your operation.

★ Chemicals are not properly labeled.

★ You do not recognize a person who is in your kitchen.

You must avoid these practices at all times to ensure that you're serving safe food. If you do not avoid these practices, people may become ill or die from consuming contaminated food from your foodservice operation.

Some other consequences of food contamination include financial devastation of the establishment, lack of consumer and governmental trust, and, potentially, the destruction of an entire global industry. Food defense is very simple: If you are aware of and pay close attention to your surroundings, fellow employees, and customers, you can help to ensure these disasters don't occur. **You** can make that difference!

Having an "It Won't Happen to Me" Attitude Can Be Very Damaging!

The attitudes of managers and employees in a food establishment can also create vulnerability when it comes to food safety. It is natural to think that nothing bad will happen to you or in your own workplace—the **"it won't happen to me"** syndrome. However, this attitude can lead to disaster. Food terrorism can potentially occur when there is

★ **Lack of awareness.** If managers and employees are not concerned about the condition of their workplace, they won't concern themselves with food defense either. The employee who thinks it is not his job to worry about food defense and the manager who thinks she provides enough food defense when it is insufficient together make the food supply of the operation more vulnerable.

★ **Limited budget for food defense measures.** Because leaders and managers have limited capital, they may not have budgeted for food defense, since it is new to the foodservice industry. This has a direct impact on the establishment's profit-and-loss statement, and additional sales must be made to recoup the difference.

★ **Lack of training.** Management's lack of knowledge about food defense may limit employees' ability to provide appropriate food defense because the team does not know or understand why it is important. It is critical to educate your employees about the **key principles of food defense—** overall awareness of your customers, fellow employees, and your surroundings. It's important to let employees know that the typical aggressor, the one wanting to contaminate the food in your establishment, relies on employees who are distracted and careless.

Your Responsibilities as a Manager

★ **Be alert and aware.** Management should make efforts to inform and involve staff in all aspects of food defense. This means you should talk about **food defense awareness** with all team members on a regular basis. An informed and alert team is more likely to detect weaknesses in a food defense system and to detect and properly respond to signs of intentional contamination. Employees should be encouraged to **report**

suspicious activities, possible product tampering, or suspected security system weaknesses to management.

★ **Encourage communication.** Communication among all employees will help to identify and correct food defense issues, raise awareness, and decrease vulnerabilities in your food defense plan. Communication also helps employees understand how important their job responsibilities are to the success of the overall operation. This realization will also help improve their food defense behaviors and attitudes.

★ **Decrease vulnerabilities.** To decrease vulnerabilities, management should develop a food defense plan. This plan should include a **recall strategy**. Staff should be **trained** on the points of the strategy and the strategy should be tested regularly, in the form of a **practical drill**. Since potential threats, the physical facility itself, and the operations within the facility may change over time, the food defense plan should be **examined and tested at least once a year**.

★ **Decrease availability of potential contaminants.** When training employees, the management team needs to stress the importance of storing all potential contaminants, such as cleaning and pest control chemicals, in **secure storage areas** that are locked. Contaminants stored on the premises should be limited to only what is needed. You should not keep a large inventory of these products. It is critical that these products are stored away from food **properly labeled, and kept in a locked area**.

Access to storage areas for these potential contaminants should be limited to those who need access, based on their job function. These storage areas should be accessed by **employees only**, and should be far away from the public areas in your establishment.

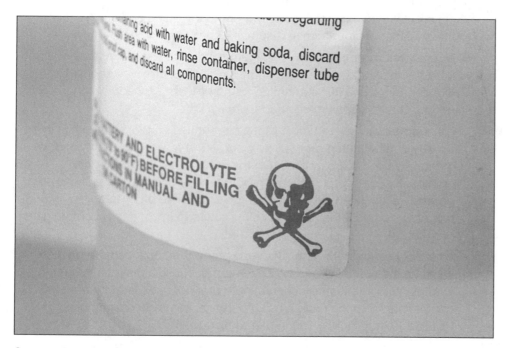

Courtesy PhotoDisc/Getty Images.

Large amounts of cleaning/chemical products add to the inventories that must be controlled and managed. **Proper inventory control** of these items helps management investigate any missing products. Additionally, any unneeded items should be properly disposed of to prevent unwanted misuse. Keep in mind: Readily available cleaning and pest control chemicals are often the contaminant of choice for a disgruntled employee who may want to get back at a manager or coworker by contaminating the food served by the establishment.

★ **Identify possible inside threats. Insider compromise** occurs when an employee (an insider) takes deliberate action to contaminate food or assists outsiders in deliberately contaminating food. In other words, it's an "inside job." The best way to prevent this from happening is to have **constant supervision** of all people working in your establishment. This includes all employees—both new and long-term employees, cleaning or maintenance staff, and vendors. Managers should be particularly aware of a person's unexplained

  ★ Hours (coming in early or staying late)

  ★ Use of camera phones, cameras, or video recorders at work

  ★ Probing with inappropriate questions

  ★ Admission to areas

  ★ Removal of standard operating procedures, documents, and policies from the facility

In addition, managers should:

★ **Conduct employee background checks.** All current and future employees, especially those with the most access to all areas of your establishment, should be screened to make sure they are not likely to perform something illegal based on their record and past behaviors. When checking employment history, follow your company SOPs and always follow the guidance of your human resource department or legal counsel.

★ **Keep track of employees.** Management should also keep track of who is and who should be on duty, and where each employee should be working by paying close attention to the prepared schedule (schedules, people charts, deployment charts, etc.).

It's important to keep in mind that a disgruntled employee who has intentionally contaminated product may not return to work and law enforcement may need to be notified. On the other hand, if a disgruntled employee plans to contaminate product, he or she may be in areas *not* normally associated with his or her job function in order to collect information or to take other actions that will help to intentionally contaminate product.

★ **Restrict personal items at the workplace.** Personal items should always be restricted in your establishment because a disgruntled employee who plans to intentionally contaminate product may need to bring the contaminant into the facility, using personal items such as a purse, backpack/knapsack, or lunch bag to disguise it.

Courtesy PhotoDisc, Inc.

Management has the authority to inspect all employee lockers and should do this periodically or if an employee is suspected of foul play. A disgruntled employee may use his or her locker as a temporary storage location for a contaminant that he or she has managed to bring into the facility. **Managers should check with their legal counsel or human resource department for proper procedures and required notifications to employees.**

Management Must Identify Possible Outside Threats

It is important for management to realize that there is also potential for an **exterior attack** or a deliberate contamination of food initiated by someone working outside your establishment. These people would include suppliers and delivery drivers.

To help avoid exterior attacks, managers should:

★ **Purchase product from approved licensed sources.** Food items should only be purchased from known and trusted sources. An unknown source may say it is a legitimate business but may offer counterfeit or contaminated food at a reduced price.

★ **Encourage suppliers to practice food defense.** Suppliers should be encouraged to practice food defense. Contamination of food can occur at a supplier's facility, and the contaminated food can then be delivered to your establishment. You should require that specific security measures are part of every supplier's contract.

★ **Inspect delivery vehicles.** Delivery vehicles should be properly inspected and secured. Locked and/or sealed vehicles keep food from being contaminated while en route to your establishment. A seal is an extra security measure that could indicate a shipment has been altered or tampered. When seals are used, compare the seal numbers on the truck or package. They should **match**. Reject the delivery if the seal numbers do not match or if the lock has been removed or if there is a potential security breach. Always err on the side of caution.

★ **Know in advance when deliveries are being made.** Management should create pickup and delivery schedules in advance. **Unscheduled pickups or deliveries should be questioned.** Delivering counterfeit or contaminated food may result in a delay of delivery, due to switching or tampering with the load and/or replacing the original driver (i.e., a hijacked load). The manager should know when a delivery is due, as well as the name of the driver, and question anything out of the ordinary.

Courtesy PhotoDisc, Inc.

★ **Supervise deliveries.** A designated employee or member of management should **supervise offloading of deliveries**. Contamination can occur during offloading, especially if it occurs after hours. The food type and quantity received should **match** what is listed on the invoice.

★ **Inspect deliveries.** A designated employee or member of management should **inspect food, packaging, and paperwork** when accepting a delivery. Those attempting to contaminate food may leave signs of contamination on the food. Such indications include abnormal powders, liquids, stains, or odors that accompany the food or evidence that tamper-resistant packaging has been resealed. Counterfeit food may contain improper or mismatched product identity, labeling, or coding. In other words, you may open a can that is supposed to contain peaches, but it contains pizza sauce instead. You should also be aware of invoices with suspicious changes, as these may accompany counterfeit or contaminated loads.

★ **Put fencing up around nonpublic areas.** Perimeter fencing should be provided for nonpublic areas of your establishment, because this is the first line of defense against attack by an intruder. The establishment should take measures to protect doors, windows, roof and vent openings, and other access points, including access to food storage (portable refrigeration units, walk-ins, tanks, and bins) outside the building(s). Locks, alarms, video surveillance, and guards can also keep intruders from breaking in.

★ **Secure access to all utilities and systems.** All utilities including gas, electric, water, and airflow systems should be secured. The greatest concern is the water supply because hazards introduced to water can be passed on to the food.

★ **Provide sufficient lighting.** Good interior and exterior lighting should be provided to deter a potential contamination. A brightly lit establishment will help prevent an attack and it will be easier to detect intruders before they contaminate any of the products.

★ **Create a system of identification.** Management should provide a means of employee identification such as a **uniform and name tag**. Requiring employees to wear identification will help to more easily identify an intruder.

★ **Restrict entry.** Entry to nonpublic areas of the establishment should be restricted. Ladders accessing the roof of your facility must be locked at all times. **Security checkpoints** should be created, and a designated employee should walk with all visitors through nonpublic areas of the establishment. You should make sure those visitors are whom they claim and that their visit has a valid purpose. This will prevent those with criminal intent from entering the establishment. Always check identification!

★ **Examine packages or briefcases.** Management should examine packages that are left in nonpublic areas, since a visitor may conceal a contaminant in a package or briefcase that he or she brings into the establishment.

★ **Monitor public areas.** Public areas should be monitored for suspicious activity. A customer who wants to contaminate food may return food that he or she has already contaminated to a shelf. Other activities include a customer spending a considerable amount of time in one part of the establishment contaminating food. Self-service areas (i.e., salad bars, product display areas, etc.), in particular, should be monitored. As mentioned earlier in this chapter, intentional contamination of a self-service location has already occurred in the United States.

★ **Use a Food Defense Checklist.** The following tool was provided by the New York State Department of Health to help you identify food defense problems that need to be corrected in your foodservice operation.

Food Defense Self-Assessment Checklist

Instructions: Use this checklist to perform a food defense assessment of your facility. After answering "Yes" or "No" to each of the questions, refer to the *Food Defense Strategies* brochure for information on how to improve food safety and security at your facility.

Facility Security
Use these questions to evaluate your facility's security.

	Yes	No	Action to take
Does management do a daily walk-through inspection of the operation?			
Is the area around the facility well lighted?			
Is the facility locked and secured when closed?			
Are exterior doors (other than customer entryways) locked at all times?			
Is access to exterior door and storage area keys restricted to management staff?			

Facility Employees
Evaluate your personnel/ access practices by answering these questions.

	Yes	No	Action to take
Are new employees' work references, addresses, phone numbers, and information on criminal record and immigration status verified?			
Is management alert for unusual employee behavior, i.e. workers staying after shift, arriving unusually early, accessing areas outside their responsibility, etc.?			
Have employees been trained in security procedures?			
Have staff been instructed to report unusual activities in the facility or on grounds?			
Is employee access restricted to those areas in the facility necessary to their job functions?			
Are customers restricted from entering food storage and preparation areas?			
Are cleaning crews, contractors or other non-facility personnel supervised while in food storage and preparation areas?			
Are employee personal items restricted to non-food handling areas?			

Receiving Supplies

Examination of your products and ingredients can prevent you from serving a problem to your customers.

	Yes	No	Action to take
Are foods purchased only from reputable vendors?			
Are deliveries received only while staff is present?			
Is the delivery person escorted while in the food storage and preparation areas?			
Are deliveries inspected for damage, tampering or counterfeiting before acceptance?			
Are delivery items matched against order invoices before acceptance/use?			
Once received, are foods immediately moved to a secure food storage area?			

Food Preparation/ Holding/ Customer Service

How food is prepared and held in an establishment has great impact on the end result, both in quality and safety.

	Yes	No	Action to take
Are standard operating procedures in place that outline the steps in each job?			
Is each ingredient and its packaging inspected for evidence of tampering before use?			
Are employees trained not to use any food with an unusual look or smell?			
Are thermometers routinely used to measure food temperatures during preparation and holding?			
Are foods thoroughly cooked to required temperatures? *(Refer to applicable Food Code)*			
Are cold foods kept below 41°F/5⁰C?			
Are hot foods kept above 135°F/57⁰C?			
Is a "no bare hand contact" policy (use of gloves or tongs) in effect and enforced with ready-to-eat items?			
Are self-service stations, like salad bars and buffets, in locations that are visible to employees at all times?			
Are empty food containers removed and replaced by new ones to replenish food at salad bars and buffets?			

Cleaning

Your cleaning and sanitizing practices help prevent contamination of your food.

	Yes	No	Action to take
Are cleaning and sanitizing chemicals used according to manufacturers' recommendations?			
Is the cleaner concentration or water temperature for dish-sanitizing routinely checked?			

REALITY CHECK

Read the scenario below and answer the questions regarding food defense concerns that relate to this situation. While reading the scenario, keep in mind if there are some actions that Doug or Zeke can take to run a safe operation.

Doug figured that his pounding headache and slight nausea were the result of rooting for the losing team in last night's Superbowl game. After drowning his sorrows in beer last night and squeezing in 4 hours of sleep, he opened the restaurant at 6:00 a.m. this morning. Not only was it a rainy morning, but he was late, so he missed the delivery truck. Since Doug wasn't at the restaurant when the driver arrived, the order was left outside in the pouring rain. Once Doug dragged the delivery inside, he quickly realized that last night's crew rushed to close the restaurant so they could all go home to watch the big game. They had done an unacceptable cleaning job; food was not put away, floors were not swept, and they forgot to "jimmy" the broken door handle on the walk-in with the screwdriver. (Doug makes a mental note to get that repaired.)

Later, Doug checked the phone messages:

★ **Message 1.** His closer from last night told him to call a plumber—all the restroom toilets are clogged.

★ **Message 2.** His assistant, Jason, who was at the same party as Doug, called to say he was very ill and couldn't come into work today.

★ **Message 3.** The printer said the new menus were all printed with the old prices and they wouldn't be ready until late next week.

Needless to say, Doug's headache was getting worse.

As the day progressed, Doug encountered disgruntled customers who had to wait too long for their food because several workers called in sick. He told his staff to do whatever it took to get the orders out. In the meantime, another delivery arrived and Doug tipped the driver to put the food away for him, since he was just too busy to check it in himself. Then, two customers complained about cold food.

Eventually, Doug begged his brother-in-law, Zeke, to come in and help. Zeke never worked in a restaurant before, but he caught on very quickly. He helped put deliveries away, served customers, bused tables, and made sandwiches.

Zeke thought it was really nice of the new pest control sales rep to leave some samples of a new product for them to test when he stopped by for a visit. He put them on the worktable so Doug could see them later. Since Zeke wasn't used to working in such a hot kitchen, he left the back door open to give everyone some relief. As he opened the door, he overheard Doug and a recently fired employee arguing outside. Zeke told a waitress to make some sandwiches so he could take them out and attempt to break up the fight. Doug would not let the hostile employee inside to get his last paycheck, but he let his girlfriend come into the office to pick it up instead.

Doug then had to excuse himself because he was still feeling ill. He returned to sign the check. The girlfriend walked through the kitchen, into the dining room, and out the front door to her waiting boyfriend. He yelled some obscenities before driving off in a rage and crashing into three parked cars, disrupting traffic.

Doug's headache was not getting any better. The police and local news crew came in for information. Apparently, the recently terminated employee had an outstanding arrest warrant from a neighboring state. Doug pleaded with the news crew to stay outside while he talked to the police privately in his office. Doug removed a strange large black duffel bag from the chair in his office and sat down to give as much information as possible. Meanwhile, the news crew was getting plenty of footage from the chaos in front of the restaurant, and wary customers who said things like "I knew something was wrong today," "The service and the food were terrible," and "This place probably does the 'breath test' and hires any one that breathes."

Doug was sure he would be feeling much worse before he started feeling any better. How was your day?

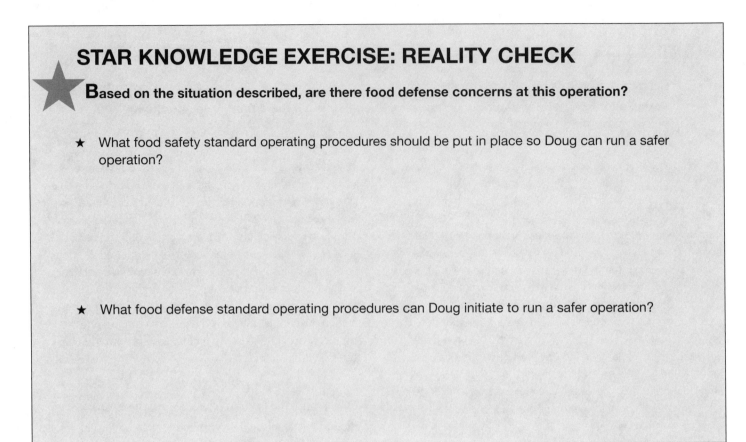

STAR KNOWLEDGE EXERCISE: REALITY CHECK

Based on the situation described, are there food defense concerns at this operation?

★ What food safety standard operating procedures should be put in place so Doug can run a safer operation?

★ What food defense standard operating procedures can Doug initiate to run a safer operation?

★ TRAINING EMPLOYEES IN FOOD DEFENSE

As a leader, you need to coach and train your employees on why it is important to prepare, serve, and sell safe food to protect you, your family, your business, and the foodservice industry. The FDA recommends that you take food defense steps to ensure the safety of your customers, your coworkers, and your country. When you, as the leader, train your employees, it's important to cover the following:

★ Employee Awareness SOP

★ Customer Awareness SOP

★ Vendor Awareness SOP

★ Facility Awareness SOP

The following checklist is provided for you to train your employees in applying food defense with a ready-to-use form that will assist you in covering all of the necessary food defense standard operating procedures.

Establishment's Name: _____

Employee Name: _____   **Employee ID#:** _____

EMPLOYEE AWARENESS SOP: Employee Initials

★ Be a responsible employee. Communicate any potential food defense issues to your manager.   ____

★ Be aware of your surroundings and pay close attention to customers and employees who are acting suspiciously.   ____

★ Limit the amount of personal items you bring into your work establishment.   ____

★ Be aware of who is working at a given time and where (in what area) they are supposed to be working.   ____

★ Periodically monitor the salad bar and food displays.   ____

★ Make sure labeled chemicals are in a designated storage area.   ____

★ Make sure you and your coworkers are following company guidelines. If you have any questions or feel as though company guidelines are not being followed, please ask your manager to assist you.   ____

★ Take all threats seriously, even if it is a fellow coworker blowing off steam about your manager and what he or she wants to do to get back at your manager or your company; or if he or she is angry and wants to harm the manager, the customer, or the business.   ____

★ If the back door is supposed to be locked and secure, **make sure it is!**   ____

★ If you use a food product every day and it is supposed to be blue but today it is green, stop using the product and notify your manager.   ____

★ If you know an employee is no longer with your company and this person enters an "employees only" area, notify your manager immediately.   ____

★ Cooperate in all investigations.   ____

★ Do not talk to the media; refer all questions to your corporate office.   ____

★ If you are aware of a hoax, notify your manager immediately.   ____

CUSTOMER AWARENESS SOP:

★ Be aware of any unattended bags or briefcases customers bring into your operation.   ____

★ If a customer walks into an "employee only" area of your operation, ask the customer politely if he or she needs help, then notify a member of management.   ____

VENDOR AWARENESS SOP:

★ Check the identification of any vendor or service person that enters restricted areas of your operation and do not leave him or her unattended.   ____

★ Monitor all products received and look for any signs of tampering.   ____

★ When a vendor is making a delivery, never accept more items than what is listed on your invoice. If the vendor attempts to give you more items than what is listed, notify your manager.   ____

★ When receiving deliveries:
   Step 1. Always ask for identification.
   Step 2. Stay with the delivery person.
   Step 3. Do not allow the person to roam freely throughout your operation.   ____

FACILITY AWARENESS SOP:	Employee Initials
★ Report all equipment, maintenance, and security issues to your manager.	_____
★ Document any equipment, maintenance, and security issues.	_____
★ Be aware of the inside and outside of your facility, including the dumpster area, and report anything out of the ordinary.	_____

ESTABLISHMENT SPECIFIC FOOD DEFENSE (OPTIONAL):

Employee Signature: Date: Owner/Manager Signature: Date:

Food Defense Tips to Remember

★ When it comes to food defense, it is not a guessing game. It is all about the facts!

★ Be aware of what to do if any acts of food terrorism or suspected hoaxes occur and always report any potential threat to your manager/supervisor or corporation and the authorities. Follow the Food Defense SOPs established by your company.

★ Remember that doing nothing at all is still taking action! By not taking action, you are allowing such hazardous situations to occur.

Whether you are experiencing issues relating to food safety, food defense, or food recalls, you should have a **crisis management plan** in place to prepare for such situations.

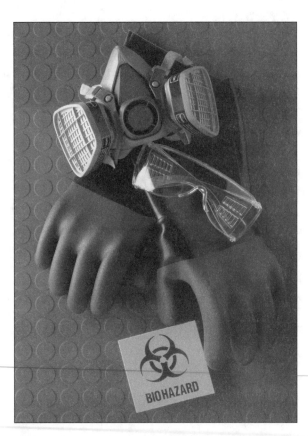

Courtesy PhotoDisc, Inc.

■ CRISIS MANAGEMENT

Many people believe that a crisis will never happen to them, that is because they never had a problem before, why should they worry now? Experts believe the question is not **whether** a crisis will occur, but **when** will it occur?

Crisis management is an organized and systematic effort to

- ★ Restrict the possibility of a likely crisis.
- ★ Manage and conclude an existing crisis.
- ★ Evaluate and learn from a crisis incident.

If you prepare for a crisis before it happens, you will be able to respond quickly and make better decisions if a crisis does indeed occur. This idea is the foundation for a HACCP plan. If you take measures to ensure food safety throughout all the steps of food preparation, ideally, you will avoid a crisis.

There are several types of crises that can be associated with the foodservice industry:

- ★ Foodborne illness complaints
- ★ Foodborne illness outbreak
- ★ Product recall
- ★ Food defense issues
- ★ Hoax
- ★ Robbery, which may involve the following:
  Theft
  Break-ins
  Murders
  Bomb threats
- ★ Natural disasters, including:
  Floods
  Earthquakes
  Electrical storms
- ★ Other crises, including:
  Fires
  Power outage
  Water failures
  Injuries

Preparing a Crisis Management Plan

It is essential that every foodservice operation develop a plan to respond to potential crises. The following examples are crisis plans for a foodborne illness crisis and a food recall crisis.

Crisis Management: Foodborne Illness Complaints

When a HACCP plan is implemented successfully, foodborne illnesses typically don't occur. However, if any part of the HACCP plan breaks down, foodborne illness may occur as a result. To respond to complaints of foodborne illness and to prevent the effects these complaints may have on the food operation's reputation, you should be sure to

★ Obtain complete, reliable information

★ Evaluate the complaint

★ Cooperate with regulatory agencies and the media

★ Reapply HACCP standards to initiate corrections, prevent recurrence, and reduce liability of the establishment

Step 1: Plan and develop policies and procedures.

The objectives of the plan should be to preserve human life, property, and supplies, and to correct the existing problems that have brought on the crisis. By planning in advance of a crisis, you will be able to respond quickly and make better and more intelligent and informed decisions.

An effective plan includes the following:

★ Identification of potential areas of crisis and steps to avoid such situations

★ Methods for management and resolution of crises

★ Evaluation during and after crises

Step 2: Designate specific person or a "crisis team."

Refer all food illness complaints to the person or team designated to respond to food-related crisis. If it is a single person, it should be the manager or someone with similar authority, while a team can include additional employees. Every employee should be trained in the policies and procedures for getting basic information from the customer lodging the complaint. They should also be instructed to refer any food-related complaints to the "crisis person" or "crisis team" for handling.

Courtesy Artville/Getty Images.

The designated crisis person or team should be given the authority to act and to direct other employees. All statements made to complainants, regulatory authorities, or the media on behalf of your food establishment should come only from the crisis person or team to ensure consistent information is being dispersed.

Step 3: Take the complaint and obtain information. Use standardized forms and procedures.

Whoever is designated to handle recording the complaints should follow these guidelines:

★ Get all pertinent information possible, without "pressuring" the complainant. Use standardized forms to avoid omitting any information.

★ Remain polite and concerned. Use your interpersonal skills. Don't argue, but don't admit liability. For example, you might say, "I'm sorry you are not feeling well," **not** "I'm sorry our food made you sick." Also, don't offer to pay medical bills or other costs, except with proper approval from your corporation or on the advice of your attorney or insurance agent.

★ Let the person tell his or her own story so you don't introduce symptoms. People tend to be suggestible about illness, and if you inquire about a symptom, they may report that symptom, because they think they should. You should only record what the person tells you.

★ Note the time the symptoms started. This is very helpful in identifying the disease and can work to clear your involvement.

★ Try to get a recent food history from the complainant. Most people blame illness on the last meal they ate, but many diseases have longer incubation periods. Again, this could clear your involvement. However, don't press—most of us can't remember what we ate for more than a few hours. If applicable, try to include food eaten before and after the person was in your establishment or during the last 48 hours.

★ Don't play doctor! Avoid the temptation to interpret symptoms or advise on treatment. Simply gather the information, remain polite and concerned, and tell the complainant you will be in touch with him or her in the near future. Then contact the complainant when you have something to report.

Foodborne Illness Complaint Form

Date_____   Time_____

Information received by (complete name) _____

Person in charge notified (complete name / date / time) _____

Complainant information (complete name) _____

Phone H (_____)_____   Cell (_____)_____   W (_____)_____

E-mail address_____   Fax_____

Mailing address_____

_____

WHO: Name(s) and contact information of people with illness_____

_____

_____

WHAT: Alleged complaint _____

_____

_____

_____

WHERE: Location_____

_____

WHEN: Date/time food was eaten_____

     Date/time of illness_____

WHY: Ask the symptoms that they are experiencing / do not assist / just LISTEN_____

_____

_____

HOW: Suspect food _____

_____

Local regulatory authorities notified (yes/no)_____

Comments_____

_____

_____

_____

_____

_____

_____

_____

_____

_____

_____

Step 4: Evaluate the complaint.

The next step will be to evaluate the complaint, so that you can handle it and respond appropriately:

★ Evaluate the complainant's attitude. If he or she is belligerent and demanding, it may be a bluff, or it could reflect a sincere feeling that you have caused damage. Note the facts on your standardized form.

★ Resist the urge to argue with or to "pay off" the complainant.

★ Analyze data to determine whether the complainant is describing a legitimate illness. Don't provide a diagnosis; however, examine the reported information for consistency:

   ★ Did the complainant eat all of the implicated serving or just a bite? Severity and duration of illness is often dose-related.

   ★ Did anyone else in the party have the same food?

   ★ Did the symptoms occur immediately?

Most illnesses require several hours of incubation before symptoms occur. Compare the complainant's meal with other meals served during the same time period. Was anything different about the complainant's meal? Was this the only complaint you received?

If this complaint is an isolated case, follow your firm's policies regarding small tokens (meal coupons, etc.) to soothe the customer and win back patronage through goodwill. Be careful to avoid admitting liability. Even in an isolated case, you must review your processes and records for any possible unsafe practices, make needed corrections, and file the complaint form for future reference.

If your evaluation indicates that the complaint is valid, obtain outside resources to help with your investigation. Contact your local health department or regulatory agency, your attorney, your media spokesperson, and your insurance agent promptly.

Whether you decide to handle the complaint privately or get involved with regulators and outside advisors, you should investigate the complaint by reapplying the principles of HACCP. First, create a flowchart to represent the activity of the implicated food. Determine whether critical control points were properly controlled. Next, review the information and your records, including the following:

★ Menus and relevant forms and logs

★ Numbers of implicated meals served, and other complaints

★ Recent changes in suppliers, employees, process, and volume

★ Correctly operating equipment

★ Recent employee illnesses (before and after the implicated meal)

★ Any indication that requirements for critical control points were not met

If you still have the implicated item, remove it from food preparation and distribution and isolate it by wrapping it securely and marking it "**Do Not Use.**" If the item is frozen, it should remain frozen. Otherwise, refrigerate it pending further instructions from a food-testing laboratory. Consult with your attorney about the effects of lab-tested samples on liability. Samples could clear you, but they could also confirm diagnosis.

The health department may request its own samples. If it does, you should take duplicate samples and arrange to have the second sample analyzed separately for comparison. For your samples, it is preferable to contract with a private laboratory to compare the results. Remember, sampling and analysis are "educated guessing" and may or may not prove anything.

Use all the combined information to make any needed changes to your operation to prevent recurrence (or use the data to clear your establishment of any wrongdoing). Resolve the complaint in accordance with your policies and advice from your attorney and insurance agent regarding the issuance of coupons or payments. Keep all complaints filed and indexed for future reference.

Health Department/Regulatory Agency as a Valued Partner

The role of regulatory agencies is to protect the consumer by helping to support and educate foodservice operators. Regulatory agencies are valued partners to foodservice operators because they provide plan reviews, third-party audits on a regular basis, and additional resources like support during a food recall or a foodborne illness outbreak, food defense checklists, self-audit checklists, and HACCP forms. Often operators don't know whom to turn to when they think they are experiencing a foodborne illness outbreak. This is why these relationships need to be built **prior** to any crisis.

Throughout the United States and the world, all operators from farm to table are encouraged to build relationships, establish communication, and participate in educational opportunities with regulators by participating in the International Association for Food Protection (www.foodprotection.org) and AFDO (Association of Food and Drug Officials; www.afdo.org). AFDO is an international nonprofit organization that has six regional affiliates that serve as resources for you and your foodservice operation:

★ CASA—Central Atlantic States Association of Food and Drug Officials; www.casafdo.org

★ AFDOSS—Association of Food and Drug Officials of the Southern States; www.afdoss.org

★ MCAFDO—Mid-Continental Association of Food and Drug Officials; www.mcafdo.org

★ NCAFDO—North Central Association of Food and Drug Officials; www.ncafdo.org

★ WAFDO—Western Association of Food and Drug Officials; www.wafdo.org

★ NEFDOA—North East Food and Drug Officials Association; www.nefdoa.org

When should you call the health department? The safest course to take is to call the health department and any other regulatory agencies (such as the Department of Agriculture or the FDA) to assist you if you believe the complaint is valid, and certainly, if there are more than two complaints or hospitalizations involved. It is to your benefit to establish a cooperative stance by involving these regulatory agencies from the start, when the complaint is first lodged against your establishment.

You should consult with your attorney about the specific laws in your jurisdiction concerning your rights and responsibilities. Generally, however, the health department is authorized to

★ Take reasonable samples of suspect foods

★ Prevent the sale of suspect foods

★ Require medical and laboratory examinations of employees

★ Exclude suspect employees from food-handling duties

★ Order the facility to be closed due to "imminent health hazards" present in your facility in extreme cases. In the 2005 Food Code the definition of **imminent health hazard** is

... a significant threat or danger to health that is considered to exist when there is evidence sufficient to show that a product, practice, circumstance, or event creates a situation that requires immediate correction or cessation of operation to prevent injury based on:

(1) The number of potential injuries, and

(2) The nature, severity, and duration of the anticipated injury.

The health department can close your establishment for the following reasons:

★ Your refrigeration is not working properly.

★ Sewage is backing up into your establishment or its water supply.

★ There is an emergency such as a flood or building fire.

★ Your establishment is suffering from a serious rodent or insect infestation.

★ Your establishment lacks electrical or water service for a long period of time.

★ Your establishment has a certain amount of critical violations.

As a manager, you should make it a point to deal effectively with the health department and other regulatory agencies. This is important because depending on the situation and your interaction, these agencies have the potential to help you and to minimize negative effects of a particular incident (including publicity), or to harm your operation by overreacting or making strong statements to the media.

An inspection by health department investigators may help to clear you in a dispute. You should always be frank and candid; don't get caught in what appears to be a "cover-up." When dealing with these investigators you should

★ Be cooperative.

★ Provide appropriate records (customer charge slips and dealer invoices) and make these records available for review.

★ Provide investigators reasonable access to observe whatever they request.

Courtesy Digital Vision.

Dealing with the Media

In today's world, technology ensures that the media and public are notified of a crisis within minutes of its being reported. The speed at which this news can travel can destroy a brand, company, or business if the information is not managed properly. You must be prepared to handle the media in case of any crisis, especially if foodborne illness occurs in your establishment.

When dealing with the media, you should be cautiously cooperative. Have your facts straight before you speak; answer simply, without jargon; and avoid creating a controversy that could make the story more interesting. Remain as professional as possible, stay calm, and don't allow yourself to be provoked. Keep your answers positive, not defensive. It is possible to use the media to your advantage, to tell your side of the story in a favorable light.

One planning technique when preparing to speak to the media suggests that you imagine your worst nightmares being described in the newspapers or on television, then practice answering those potentially embarrassing questions until you feel and act comfortable.

Stick to the truth and don't try to bluff. If you don't know the answer, admit it and arrange to respond to the question at a later time, after you have investigated the question further. Also, remember that "no comment" tends to sound like an admission of guilt. Managers should remind employees that they are not to talk to the media; all requests made of employees should be referred to management or the corporation. Here is a sample model press release with example provided by the FDA ORA/Office of Enforcement Division of Compliance Management and Operations (www.fda.gov/ora/compliance_ref/recalls/ecoli.htm).

E. coli 0157:H7 Model Press Release

FOR IMMEDIATE RELEASE   **DATE**
COMPANY CONTACT AND PHONE NUMBER

FOOD CO. RECALLS **PRODUCT** BECAUSE OF POSSIBLE HEALTH RISK

Company Name of **City, State** is recalling **Quantity and/or Type of Product** because it may be contaminated with Escherichia coli 0157:H7 bacteria (E. Coli 0157:H7). E. coli 0157:H7 causes a diarrheal illness often with bloody stools. Although most healthy adults can recover completely within a week, some people can develop a form of kidney failure called Hemolytic Uremic Syndrome (HUS). HUS is most likely to occur in young children and the elderly. The condition can lead to serious kidney damage and even death.

Product was distributed by _____. Listing of states and areas where the product was distributed and how it reached consumers (e.g., through retail stores, mail order, direct delivery).

Specific information on how the product can be identified (e.g., type of container [plastic/metal/glass], size or appearance of product, product brand name, flavor, codes, expiration dates, etc.).

Status of the number of and types of related illnesses that have been CONFIRMED to date (e.g., "No illnesses have been reported to date.")

Brief explanation about what is known about the problem, such as how it was revealed, and what is known about its source. An example of such a description: "The recall was initiated after it was discovered that product was contaminated with E. coli 0157:H7. Subsequent investigation indicates the problem was caused by a temporary breakdown in the company's production and packaging processes."

Information on what consumers should do with the product and where they can get additional information (e.g., "Consumers who have purchased Brand X are urged to return it to the place of purchase for a full refund. Consumers with questions may contact the company at 1-800-XXX-XXXX").

(SAMPLE PRESS RELEASE)

XYZ Inc.
123 Smith Lane
Anywhere, MS

FOR IMMEDIATE RELEASE   **DATE**
Sam Smith /555-555-5555

XYZ RECALLS "SNACKIES" BECAUSE OF POSSIBLE HEALTH RISK

XYZ Inc. of Anywhere, MS, is recalling its 5-ounce packages of "Snackies" food treats because they have the potential to be contaminated with Escherichia coli 0157:H7. E. coli 0157:H7 causes a diarrheal illness, often with bloody stools. Although most healthy adults can recover completely within a week, some people can develop a form of kidney failure called Hemolytic Uremic Syndrome (HUS). HUS is most likely to occur in young children and the elderly; the condition can lead to serious kidney damage and even death.

The recalled "Snackies" were distributed nationwide in retail stores and through mail orders.

The product comes in a 5-ounce, clear-plastic package marked with lot # 7777777 on the top and with an expiration date of 03/17/06 stamped on the side.

No illnesses have been reported to date in connection with this problem.

The potential for contamination was noted after routine testing by the company detected the presence of E. coli 0157:H7.

Production of the product has been suspended while FDA and the company continue their investigation as to the cause of the problem.

Consumers who have purchased 5-ounce packages of "Snackies" are urged to return them to the place of purchase for a full refund. Consumers with questions may contact the company at 1-800-XXX-XXXX.

Crisis Management: Food Recalls

A **food recall**, as defined by the FDA, is "an action by a manufacturer or distributor to remove a food product from the market because it may cause health problems or possible death."

Recall Classification General Definition and Example

★ **Class I.** This classification involves a health hazard situation where there is a **reasonable** probability that consuming the product will cause adverse health problems or death. For example, a Class I would be a ready-to-eat product such as luncheon meat or hot dogs contaminated with a pathogen or bacteria.

★ **Class II.** This involves a potential health hazard situation where there is a **remote** probability of adverse health problems from consuming the product. A Class II example would be a product found to have small pieces of plastic in it. Another example is a product that is found to contain an allergen, such as dry milk, that is not mentioned on the label.

★ **Class III.** This involves a situation when consuming the product will not cause adverse health problems. For example, a product that has a minor ingredient missing from the label that is not an allergen, such as labeled processed meat in which added water is not listed on the label as required by federal regulations.

★ **Withdrawal.** Only the FDA uses this classification. The situation occurs when a product has a minor violation that would not be subject to FDA legal action. The company removes the product from the market or corrects the violation. A withdrawal example would be a product removed from the market due to tampering, without evidence of manufacturing or the distribution problems.

★ **Hold.** A time period used for investigation after a USDA commodity food has been identified as potentially unsafe. The hold time for commodity foods is no longer than 10 days.

★ **Release.** When the product on hold has been found safe and can be used.

For more information on recall classifications, visit:

★ www.fsis.usda.gov/OA/pubs/recallfocus.htm
★ www.cfsan.fda.gov/~lrd/recall2.html

Remember: A food recall requires immediate action! There is a limited amount of time to perform the tasks necessary to guard the safety of customers when a food recall occurs. It is best to have a standard operating procedure in place before being notified of a food recall.

Food Recall SOP

Certain steps must be taken in the event of a food recall:

1. **Develop a SOP before being notified of a food recall.** The standard operating procedure must include a list of steps that should be taken, those who are responsible for each step, and detailed procedures to be followed at each step. For the procedures to be effective, personnel must be trained to use the hold and recall process.

SOP: Food Recalls (Sample)

Date issued:

Revision: *Number or letter designation*

Purpose: To identify the procedures and personnel responsible in the event of a food recall and to document the procedure to be followed

Scope: This procedure applies to (*List **all personnel** by job title to whom this applies. Listing by title ensures that the procedure does not have to be revised and approved every time there is a personnel change.*)

General: (Definitions and general statements)

★ Recall—An action by a manufacturer or distributor to remove a food product from the market because it may cause health problems or possible death.

★ Hold—A time period used for investigation after a USDA commodity food has been identified as potentially unsafe. The hold process is unique to USDA commodity foods.

★ Release—When the product on hold has been found safe and can be used.

★ Physical segregation—Product is removed to a separate area of storage from other foods.

Personnel Responsible: (List by title for each responsibility below.)

Distributor contact _____

Documentation of training _____

Food safety coordinator _____

Public communications contact person _____

Training on recall procedures _____

Procedures: Responsibilities When a Food Recall Notice Is Received:

Personnel responsible will complete the assigned tasks described in the attached Food Recall Action Checklist. (*Add job titles to Food Recall Action Checklist under Person Responsible and attach the checklist to the standard operating procedure.*)

Communication:

Communication will be handled as follows:
(*Example: Initial communication when a Food Recall Notice is received will be phoned directly to the contact person responsible at the site. A log will be maintained to document contacts by date and time. Any printed materials needed at the sites such as recall notices will be faxed to each site.*)

Contacts for Public Communications:

Contact person will handle all public communications. (*Include the contact person or other personnel responsible for public communications. Attach a copy of the policy and procedure for public communications.*)

Product Segregation:

The recalled food product will be physically separated by: *(Describe the procedure to segregate the recalled product. The procedure may be product-type-specific. For example, frozen product placed on hold will be placed in a plastic bag, securely taped, and placed on a separate shelf.)*

Warning:

Product placed in hold will be identified by:
*(Describe the product identifier: A large sign, not less than 8-1/2 × 11 inches, with the words "***DO NOT USE***" and "***DO NOT DISCARD***" will be securely attached to the product placed on hold.)*

Requirements for On-Site Destruction of Recalled Product:

(Describe specific state or local requirements for disposal of food products. Some state public health departments require notification of all food products to be destroyed before any action is taken).

(Procedures need to be approved by at least two persons.)

Approved _____   Date _____

Approved _____   Date _____

To ensure that the Food Recall SOP will be seamless if a food recall should occur, the manager should

★ Conduct a mock food recall to assist in teaching employees the appropriate steps within this particular process.

★ Appoint a responsible person to coordinate food safety in the establishment. The manager also appoints a backup food safety coordinator in case the principle food safety coordinator is not available in the time of crisis.

Again, take action immediately when you receive notice of a recall!

2. **Review the recall notification report when it is received.**

★ Determine what the problem is according to the recall.

★ Review all instructions issued in the recall.

★ Determine what actions must be taken.

★ Gather recall information following the FDA form #3177 (www.fda.gov/ora/inspect_ref/iom/pdf/Chapter8.pdf)

The following form will be generated and completed by the FDA. In the event of a food recall, the establishments affected will receive a recall notification report from the State Distributing Agency. This report will state the name of the product, affected lot numbers, and any other information about the product. Establishments affected must return information to the State Distributing Agency. This information will include where the product listed on the recall has been stored, the quantity being stored, the quantity of the product that was already used, and documentation of the received product so that those involved can be reimbursed. The following report is the result of the initial recall investigation. Foodservice operations will take actions based on the information given to them from this report.

FORM FDA 3177

1. RECALL INFORMATION	2. PROGRAM DATA (CHECK BOX IF PREVIOUSLY SUBMITTED) (DO NOT COMPLETE IF REPORTED UNDER FDA 2123)

1. RECALL INFORMATION

a. RECALL NUMBER

b. RECALLING ESTABLISHMENT

c. RECALLED CODE(S)   d. PRODUCT

2. PROGRAM DATA (CHECK BOX IF PREVIOUSLY SUBMITTED)
(DO NOT COMPLETE IF REPORTED UNDER FDA 2123)

a. ACCOMP DISTRICT CODE	b. HOME DISTRICT CODE	c. OPERATION CODE	d. OPERATION DATE (MM/DD/YY)
		17	

e. CENTRAL FILE NUMBER OF RECALLING ESTABLISHMENT   f. PAC CODE

g. EMPLOYEE			h. TYPE	# OF CHECKS	HOURS
HOME DIST.	POS. CLASS	NUMBER	VISITS		
			PHONE		

3. AUDIT ACCOUNTS

a. DIRECT	b. SUB-ACCOUNT (SECONDARY)	c. SUB-ACCOUNT (TERTIARY)
PHONE NO. _____	PHONE NO. _____	PHONE NO. _____

4. CONSIGNEE DATA   Contacted by: ☐ Phone ☐ Visit ☐ Other

a. NAME OF PERSON CONTACTED, TITLE & DATE

b. TYPE CONSIGNEE

☐ Wholesaler   ☐ Physician
☐ Retailer   ☐ Hospital   ☐ Other
☐ Processor   ☐ Pharmacy
☐ Consumer   ☐ Restaurant

c. DOES (DID) THE CONSIGNEE HANDLE RECALLED PRODUCT?

☐ YES   ☐ NO

5. NOTIFICATION DATA

a. FORMAL RECALL NOTICE RECEIVED? *(If "No" skip to item 6c.)*

☐ YES   ☐ NO   ☐ CANNOT BE DETERMINED

b. RECALL NOTIFICATION RECEIVED FROM:

☐ Recalling Firm
☐ Direct Account
☐ Sub-Account
☐ Other *(Specify)*

c. DATE NOTIFIED

d. TYPE OF NOTICE RECEIVED (e.g. letter, phone)

6. ACTION AND STATUS DATA

a. DID CONSIGNEE FOLLOW THE RECALL INSTRUCTIONS? *(If "No", discuss in item 10 action taken upon FDA contact)*
☐ YES   ☐ NO

b. AMOUNT OF RECALLED PRODUCT ON HAND AT TIME OF NOTIFICATION

c. CURRENT STATUS OF RECALLED ITEMS
☐ Returned   ☐ Destroyed
☐ Corrected   ☐ None on Hand
☐ Was Still Held for Sale/Use *
☐ Held For Return/Correction *
* = Ensure Proper Quarantine/Action

d. DATE AND METHOD OF DISPOSITION

7. SUB-RECALL NEEDED?
Did Consignee Distribute to any other Accounts?
(If "Yes" give Details in "Remarks" or Memo)
☐ YES   ☐ NO

8. AMOUNT OF RECALLED PRODUCT NOW ON HAND

9. INJURIES/COMPLAINTS

IS CONSIGNEE AWARE OF ANY INJURIES, ILLNESS, OR COMPLAINTS?

☐ INJURY   ☐ COMPLAINT
☐ ILLNESS   ☐ NONE

If answer is other than "None", report details in a separate memo to monitoring district and copy to E.O.B. (HFC-162)

10. REMARKS *(Include action taken if product was still available for sale or use)*

SIGNATURE OF CSO/CSI

TO: DATE   ENDORSEMENT

DISTRICT   DATE OF CHECK   SIGNATURE OF SCSO OR R&E COORDINATOR

Sample Food Recall Notification Report from the USDA Food Safety and Inspection Service (www.fsis.usda.gov/oa/recalls/rnrfiles/rnr090-2002.htm)

October 13, 2002

Recall Notification Report
U.S. DEPARTMENT OF AGRICULTURE
FOOD SAFETY AND INSPECTION SERVICE

EXPANDED

Product(s) Recalled:	Turkey and Chicken Products On October 03, 2002, and October 04, 2002, FSIS collected samples from structural and equipment surfaces at the establishment as part of an ongoing food safety investigation. Several of the samples were reported positive for *Listeria monocytogenes* by the FSIS Midwestern Laboratory. The results strongly suggest that the general processing environment at the establishment may be the source of widespread product contamination. Based on this and following a scientific and technical review of plant practices and company records by FSIS, the firm has voluntarily expanded the recall 090-2002 to include turkey and chicken products produced between May 01, 2002, through October 11, 2002, because these products may be contaminated with *Listeria monocytogenes*. In addition, the establishment has voluntarily suspended operations.
Production Dates/Identifying Codes:	The following turkey and chicken products are subject to the expanded recall that were produced between May 01, 2002 and October 11, 2002: This list has been modified and does not include the complete listing of products recalled. • Various sized boxes of "WAMPLER FOODS OVEN ROASTED TURKEY BREAST WITH BROTH BONELESS, 10003." Packed in each box are 9-pound package of "WAMPLER FOODS TURKEY BREAST WITH BROTH • WHITE MEAT ADDED • BONELESS OVEN ROASTED 95% FAT FREE." The products subject to recall bear a sell-by date of "7/22/02 to 1/2/03." • Various sized boxes of "WAMPLER FOODS OVEN ROASTED TURKEY BREAST WITH BROTH BONELESS, 10004." Packed in each box are 9-pound packages of "WAMPLER FOODS TURKEY BREAST WITH BROTH BONELESS OVEN ROASTED 97% FAT FREE." The products subject to recall bear a sell-by date of "7/22/02 to 1/2/03." • Various sized boxes of "WAMPLER FOODS OVEN ROASTED TURKEY BREAST WITH BROTH BONELESS, 10005." Packed in each box are 9-pound packages of "WAMPLER FOODS TURKEY BREAST WITH BROTH BONELESS • OVEN ROASTED 98% FAT FREE." The products subject to recall bear a sell-by date of "7/22/02 to 1/2/03." • Various sized boxes containing 9-pound packages of "FULLY COOKED DELI STYLE SMOKED TURKEY BREAST SMOKE FLAVOR ADDED, 10006." The products subject to recall bear a code "MFG 4/30/02 to 10/10/02." • Various sized boxes containing 9-pound packages of "FULLY COOKED, DELI STYLE, TURKEY BREAST, 10007." The products subject to recall bear a code "MFG 4/30/02 to 10/10/02." All of the products bear the establishment number "P-1351" inside the USDA mark of inspection unless otherwise noted.
Problem/Reason for Recall:	The products may be contaminated with *Listeria monocytogenes*.
How/When Discovered:	In response to a food safety investigation, FSIS collected a microbiological investigative sample on October 02, 2002, that returned positive results for *Listeria monocytogenes* on October 08, 2002. On October 03, 2002, and October 04, 2002, FSIS collected samples from structural and equipment surfaces at the establishment as part of an ongoing food safety investigation. Several of the samples were reported positive for *Listeria monocytogenes* by the FSIS Midwestern Laboratory. The results strongly suggest that the general processing environment at the establishment may be the source of widespread product contamination. Based on this and following a scientific and technical review of plant practices and company records by FSIS, the firm has voluntarily expanded the recall 090-2002 to include turkey and chicken products produced between May 01, 2002, and October 11, 2002. In addition, the establishment has voluntarily suspended operations.

Federal Establishment:	01351 P Pilgrim's Pride Corporation Doing business as: Wampler Foods, Inc. 471 Harleysville Pike Franconia, PA 18924
Consumer Contact:	Consumer Information Recall Hotline, 877-260-7110
Media Contact:	Ray Atkinson, Public Relations Manager, 540-896-0406
Quantity Recalled:	Approximately **28 million pounds** (This includes the 295,000 pounds of the October 09, 2002, Recall).
Distribution:	Nationwide
Recall Classification:	Class I
Recall Notification Level:	Consumer and User (food service)
Press Release:	Yes
Direct Notification Means:	The firm has notified its customers orally and will follow up in writing.
FSIS Follow-up Activities:	Effectiveness checks by the FSIS.
Other Agencies Involved:	FSIS is working with the Centers for Disease Control and Prevention (CDC) and various northeastern state health officials.
FSIS Contacts:	• Compliance/Recall Coordinator: 202-418-8874 • Recall Management Division: 202-690-6389 • Media Inquiries: 202-720-9113 • Congressional Inquiries: 202-720-3897 • Consumer Inquiries: 1-800-535-4555 • Web Site: www.fsis.usda.gov/ (FSIS Main Page) or www.fsis.usda.gov/oa/recalls/rec_intr.htm (Recall Information Center)
Date of Recall Meeting:	October 09, 2002, **expanded on October 12, 2002**
Recall Case Number:	090-2002

For Further Information, Contact:

- **Consumers**: Meat and Poultry Hotline, 1-800-535-4555 or (202) 720-3333 (voice); 1-800-256-7072 (TTY)
- **Media:** (202) 720-3897

Source: USDA Food Safety and Inspection Service

3. Communicate information about the food recall immediately.

The manager of the food establishment must communicate the information to everyone involved in the recall. The manager must also do the following:

★ Clarify all issues related to the food recall by speaking directly to the person responsible for the recall.

★ Document and confirm receipt of all communication.

★ Immediately notify all sites of the recall.

★ Identify the location of all products affected by the recall.

★ Verify that the food items have the product identification codes listed in the recall.

★ Isolate the food products to avoid accidental use.

★ Take an accurate inventory by location.

Courtesy Corbis Digital Stock.

4. **Collect health-related information needed for public communications.**

Recall classification is determined by the seriousness of the health risks involved. For Class I or Class II recalls, the following should be documented:

★ If any of the affected food was served

★ To whom it was served

★ The dates it was served

★ Any health problems reported that could be related to the recalled food product

5. **Work closely with all public communications contact persons.**

Notify the communications contact person/media spokesperson associated with your establishment so that this person is prepared to handle all public communications with health agencies and local media. Provide the public communications contact person with all the correct information you have so he or she is well-informed of the situation, including:

★ Copies of the food recall notice

★ Press releases

★ Any related recall information

★ Information on whether the product has been used and served, to whom it has been served, and on what dates it has been served

★ Any health risk reports related to the recalled product

The extent of the information communicated to the public will depend on the type of recall and whether or not any of the food products has been consumed. If the food product has been served to customers and there is the risk of an adverse health effect, the persons potentially affected should (**ethically and legally**) be notified. If there is no potential adverse health risk associated with serving the product or if the product has not been served, there is no reason to report the recall to the customer.

6. **Locate the recalled food product.**

   ★ Determine where any of the affected food product listed in the recall notice is located.

   ★ Make sure the codes on the affected product match the code(s) listed in the recall notice.

7. **Count the inventory of the recalled food product.**

   Take an accurate inventory of the recalled food product already used and an accurate inventory of how much of the recalled food is still being stored.

8. **Account for all of the recalled food.**

   Check all records to verify that all recalled food has been accounted for and has been removed from potential use.

9. **Separate and secure the recalled food product from that food not listed on the recall.**

   Keeping the affected food separate from food not listed on the recall will prevent instances of cross-contamination from occurring.

10. **Take action to conform to the recall.**

    ★ To receive reimbursement or replacement of the recall product, and to have the affected product removed from your establishment, you must submit information to the manufacturer, distributor, or state agency. This information must describe the quantity of the product in stock and where it has been stored. The information must be submitted within 10 days of the first notification of the recall.

    ★ Find out if the product should be returned to someone, or, if the product needs to be destroyed, find out who is responsible for destroying it and how it is to be destroyed. Note that in nearly all food recalls, the distributor or vendor will collect and/or destroy the recalled food products.

    ★ Determine which employees will be responsible for obtaining this information and following the standard operating procedures involved with such actions.

    ★ Notify all parties involved of the procedures, dates, and so on to be followed in the recall or destruction of the recalled food product.

    ★ Work with the state distributing agency and the contracted warehouse/distributor to determine how and when to collect the recalled food product. The recalled product should be collected as soon as possible, and no later than 30 days after the date of the recall notice.

    ★ Do not destroy any product without official written notification from USDA (United States Department of Agriculture), FSIS (Food Safety and Inspection Service), or the FDA (United State Food and Drug Administration). Once notified by one of these agencies, if you destroy the food on-site, you must determine how the food should be destroyed, who should be there, and who must know about the process. Due to the potential health risk to humans or animals, some state public health departments require notification of all recalled food products to be destroyed before any action is taken.

11. **Consolidate documentation from all sites for inventory counts.**

    This will help the manager/director account for recalled product that has been served versus what is still in the foodservice operation.

12. **Document any reimbursable costs.**

Determine what recalled food items you should be reimbursed for by the manufacturers and/or distributors, and submit the necessary paperwork for reimbursement of food costs.

Complete and maintain all necessary documentation related to the recall. The FDA recommends keeping records for 3 years plus the current year. These examples of crisis management plans have been provided to help prepare you for such crises in your foodservice operation. Keep in mind: If you practice the seven HACCP principles, hopefully, you will never have to use a crisis management plan, since you're practicing food safety. However, it's extremely important to have a crisis management plan in place just in case a crisis occurs at your establishment. The chart provided by the U.S. Department of Agriculture assists in the managing of crisis follow-up items.

Monitoring Schedule for Items that Require Follow-up Action					
Task to monitor or follow-up	When will follow-up begin?	How often will follow-up occur?	Who is responsible?	Results of follow-up	Check when completed
					❑
					❑
					❑
					❑

Resources

We recommend that all foodservice operations should give this resource list, their employee contact list, and the completed first-responder emergency contact list to all management and the company's media spokesperson.

FDA: 1-888-INFO-FDA/1-888-463-6332 or 301-443-1240

- ★ www.fda.gov
- ★ www.fda.gov/oc/bioterrorism/bioact.html
- ★ www.fda.gov/oc/bioterrorism/report_adulteration.html
- ★ www.fda.gov/ora/training/orau/FoodSecurity/default.htm

Center for Safety and Applied Nutrition: 1-888-SAFEFOOD

- ★ www.cfsan.fda.gov/~dms/foodcode.html
- ★ www.cfsan.fda.gov/~dms/secgui11.html
- ★ www.cfsan.fda.gov/~dms/secguid6.html

FSIS/Food Safety and Inspection Service: 1-800-333-1284

- ★ www.fsis.usda.gov/oa/topics/securityguide.htm

U.S. Department of Agriculture: 202-720-3631 or 1-800-233-3935

- ★ schoolmeals.nal.usda.gov/Safety/biosecurity.pdf
- ★ www.nfsmi.org/Information/e-readinessguide.pdf
- ★ www.nfsmi.org/Information/recallmanual.pdf

Department of Homeland Security: 202-324-0001

- ★ www.ready.gov *or* http://www.dhs.gov

FBI / Federal Bureau of Investigations: 202-456-1111

- ★ www.fbi.gov

Centers for Disease Control and Prevention: 800-CDC-INFO or 888-232-6348

- ★ www.bt.cdc.gov/

American State Health Officials: 202-371-9090

- ★ www.statepublichealth.org

General Accounting Office: 202-512-4800

- ★ www.gao.gov/new.items/rc00003.pdf

First Responder Emergency Contact Information

Date compiled/updated: _____

Poison Control
Contact name: _____
Phone number: _____
Fax number: _____
Email: _____
Address: _____
_____
_____
Call if: _____
_____

Owner/Director/Person in charge
Contact name: _____
Phone number: _____
Fax number: _____
Email: _____
Address: _____
_____
_____
Call if: _____
_____

Attorney/Legal Representative
Contact name: _____
Phone number: _____
Fax number: _____
Email: _____
Address: _____
_____
_____
Call if: _____
_____

Manager
Contact name: _____
Phone number: _____
Fax number: _____
Email: _____
Address: _____
_____
_____
Call if: _____
_____

Media Spokesperson
Contact name: _____
Phone number: _____
Fax number: _____
Email: _____
Address: _____
_____
_____
Call if: _____
_____

Manager
Contact name: _____
Phone number: _____
Fax number: _____
Email: _____
Address: _____
_____
_____
Call if: _____
_____

Consultant
Contact name: _____
Phone number: _____
Fax number: _____
Email: _____
Address: _____
_____
_____
Call if: _____
_____

Manager
Contact name: _____
Phone number: _____
Fax number: _____
Email: _____
Address: _____
_____
_____
Call if: _____
_____

Attach Employee Contact List

ARE YOU A FOOD DEFENSE "SUPERSTAR"?

★ Should you get involved if you notice something or someone suspicious at your establishment? Yes or No?

★ Should you do something about the potential problem? Yes or No?

If you answered No, how well do you think you will sleep tonight if someone gets ill because you were scared or you ignored the situation?

★ **Doing NOTHING at all IS taking ACTION! By not taking any action, you are allowing such situations to occur.**

In the following situations, what can you do to provide food defense for your customers, your coworkers, your country, and the business you represent?

Situation/Scenario	Is the situation/scenario a concern? Yes or No?	If so, what do you do to fix the problem?
1. A newly hired cook approaches you and tells you he saw your best cook contaminating food.		
2. A customer wanders into an "employee only" area and you realize it is the ex-boyfriend of one of your employees.		
3. An employee tells you that the employee you wrote up yesterday is upset and has started talking about ways in which he is going to harm you and the company.		
4. A customer brings a large duffel bag into your operation near closing time. He uses the bathroom, and then waits outside, even after you close.		

Situation/Scenario	Is the situation/scenario a concern? Yes or No?	If so, what do you do to fix the problem?
5. An ex-employee returns only one uniform after she had been terminated. Your paperwork indicates she had been given two uniforms.		
6. An incident happened in your operation causing many customers to become seriously ill. A customer approaches you and starts asking very specific questions about the incident.		
7. A vendor is late making a delivery. You had ordered seven units of a certain food item, but the vendor wants to give you an extra one to make up for the inconvenience of his lateness.		
8. You walk by the food bar and see a spray bottle of glass cleaner sitting on the food bar. It isn't a bottle you recognize as being from your establishment.		
9. A vendor who has been delivering to you for years walks through the back door with a delivery of raw chicken wings. You notice the cases are dripping an odd liquid.		
10. An employee tells you there are small holes in every one of the cans of food needed to prepare meals for that day. The cans had just been delivered earlier that morning.		

In Star Point 1, we covered the need for good retail practices, active managerial control, and prerequisite programs in great detail. Now, after you have read through Star Point 2, it is clear that in addition to food safety SOPs, you also need food defense SOPs to prevent a deliberate act of food contamination from occurring. Whether it is food defense or food safety, your establishment needs standard operating procedures for both practices before your organization begins to implement the first HACCP principle: conducting a hazard analysis.

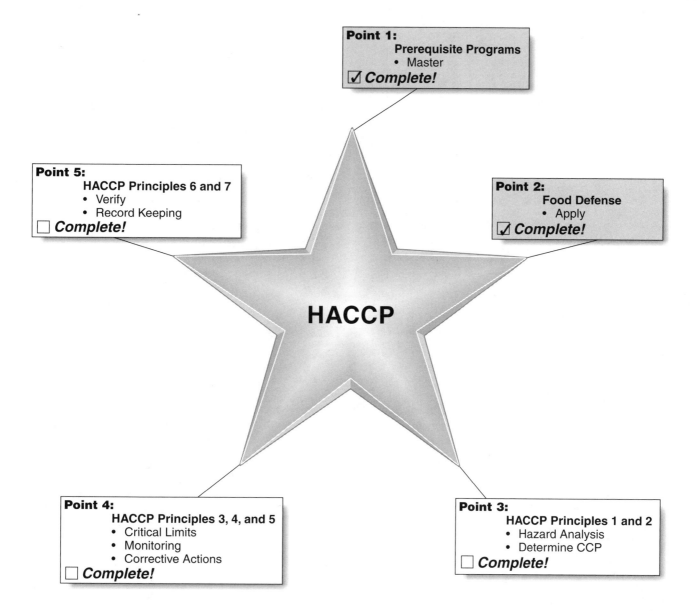

Point 1:
Prerequisite Programs
• Master
☑ *Complete!*

Point 5:
HACCP Principles 6 and 7
• Verify
• Record Keeping
☐ *Complete!*

Point 2:
Food Defense
• Apply
☑ *Complete!*

HACCP

Point 4:
HACCP Principles 3, 4, and 5
• Critical Limits
• Monitoring
• Corrective Actions
☐ *Complete!*

Point 3:
HACCP Principles 1 and 2
• Hazard Analysis
• Determine CCP
☐ *Complete!*

Create a HACCP Plan

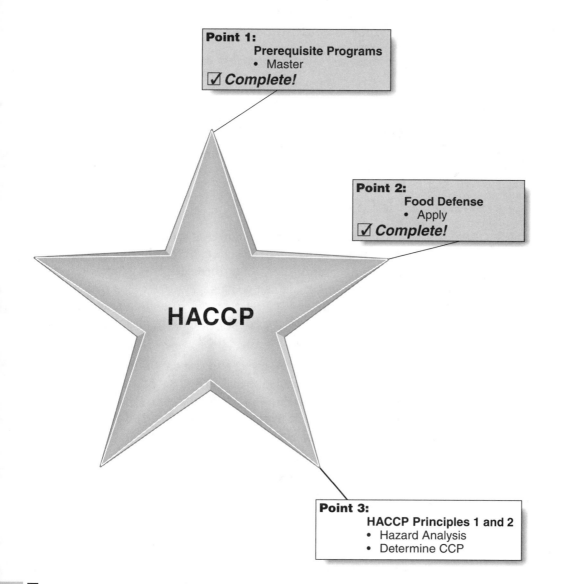

Point 1:
Prerequisite Programs
- Master
☑ *Complete!*

Point 2:
Food Defense
- Apply
☑ *Complete!*

HACCP

Point 3:
HACCP Principles 1 and 2
- Hazard Analysis
- Determine CCP

This third point of the HACCP Star examines HACCP Principles 1 (Conducting a Hazard Analysis) and 2 (Determining Critical Control Points). Now that the prerequisite programs for food safety and food defense have been established for your operation, you have the foundation to create an effective HACCP plan. Remember, however, training and monitoring employees in food safety and food defense basics is **vital** to making HACCP successful. This means that each employee must do his or her part to understand and maintain basics such as facility design, equipment, pest control, chemical control, cleaning and sanitizing procedures, responsible employee practices, purchasing practices, food specifications, and food temperature control. This Star Point also addresses the importance and philosophy of HACCP, as well as the international implications and the **Codex Alimentarius Commission**.

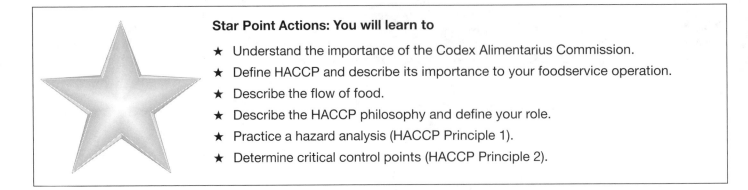

Star Point Actions: You will learn to

★ Understand the importance of the Codex Alimentarius Commission.

★ Define HACCP and describe its importance to your foodservice operation.

★ Describe the flow of food.

★ Describe the HACCP philosophy and define your role.

★ Practice a hazard analysis (HACCP Principle 1).

★ Determine critical control points (HACCP Principle 2).

★ HACCP INTRODUCTION

■ WHAT IS HACCP?

HACCP (pronounced "has-sip") is a food safety system that prevents disasters, such as foodborne illness outbreaks, from occurring. A **foodborne illness** occurs when you eat food and it makes you sick. An **outbreak** occurs when two or more people eat the same food and get the same illness. HACCP helps prevent foodborne illness outbreaks because HACCP is a proactive approach to controlling every step in the flow of food. Simply stated, the HACCP goal is to stop, control, and prevent problems that impact the safety of food.

HACCP is an abbreviation for **H**azard **A**nalysis and **C**ritical **C**ontrol **P**oint. HACCP is a written food safety system to enable the selling and serving of safe food. Star Point 1 (prerequisites programs) and Star Point 2 (food defense) have laid both the foundation for HACCP and the next three Star Points. The HACCP food safety system has seven principles. You will learn about these seven principles in the next three Star Point sections.

Star Point 3—Create the HACCP Plan:

    ★ HACCP Principle 1: Conducting a Hazard Analysis

    ★ HACCP Principle 2: Determining Critical Control Points

Star Point 4—Work the Plan:

    ★ HACCP Principle 3: Establish Critical Limits

    ★ HACCP Principle 4: Monitoring

    ★ HACCP Principle 5: Taking Corrective Actions

Star Point 5—Checks and Balances:

    ★ HACCP Principle 6: Verification

    ★ HACCP Principle 7: Record Keeping and Documentation

No matter what your job description and duties are, you will benefit from understanding and knowing the HACCP system. For many operations, like food manufacturers and public-school systems, regulatory authorities have mandated that a HACCP plan be in place at all times. Is your operation required by law to have a HACCP plan?

■ WHY IS HACCP IMPORTANT?

Based on the combined efforts of the Food and Agriculture Organization (FAO) concentrating its efforts on global nutrition, and the World Health Organization (WHO) focusing on preventing disease and promoting general well-being, the United Nations established the **Codex Alimentarius Commission** in 1963 to provide worldwide food safety and nutrition. In Latin, *Codex Alimentarius* means food law or code. The Codex Alimentarius Commission provides the world with international **food safety standardization,** which allows the consumer to be better protected and to enjoy more varieties of food products, while promoting healthy and safe choices. The Codex Alimentarius Commission is instrumental in developing and coordinating international food safety guidelines, procedures, and standards to protect the health and safety of consumers and in encouraging fair international trade around the world. The Codex Alimentarius Commission has more than 160 countries as members. These members represent 97 percent of the world population that recognizes the importance of HACCP. Following is a snapshot of the countries impacted by the power of the Codex Alimentarius Commission to facilitate international food safety. Also listed are the countries who represent Codex Alimentarius Commission Members:

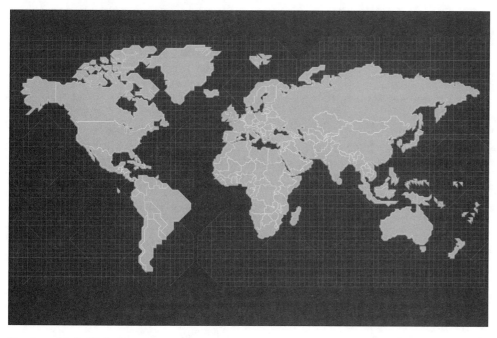

Courtesy Artville/Getty Images.

Afghanistan
Albania
Algeria
Angola
Antigua and Barbuda
Argentina
Armenia
Australia
Austria

Bahamas
Bahrain
Bangladesh
Barbados
Belgium
Belize
Benin
Bhutan
Bolivia
Botswana
Brazil
Brunei Darussalam
Bulgaria
Burkina Faso
Burundi

Cambodia
Cameroon
Canada
Cape Verde
Central African Republic
Chad
Chile
China
Colombia
Congo
Cook Islands
Costa Rica
Côte d'Ivoire
Croatia
Cuba
Cyprus
Czech Republic

Democratic People's Republic of Korea
Democratic Republic of the Congo
Denmark
Dominica
Dominican Republic

Ecuador
Egypt
El Salvador
Equatorial Guinea
Eritrea
Estonia
Ethiopia
European Community

Fiji
Finland
France

Gabon
Gambia

Georgia
Germany
Ghana
Greece
Grenada
Guatemala
Guinea
Guinea-Bissau
Guyana

Haiti
Honduras
Hungary

Iceland
India
Indonesia
Iran (Islamic Republic of)
Iraq
Ireland
Israel
Italy

Jamaica
Japan
Jordan

Kazakhstan
Kenya
Kiribati
Kuwait
Kyrgyzstan

Lao People's Democratic Republic
Latvia
Lebanon
Lesotho
Liberia
Libyan Arab Jamahiriya
Lithuania
Luxembourg

Madagascar
Malawi
Malaysia
Mali
Malta
Mauritania
Mauritius
Mexico
Micronesia (Federated States of)
Mongolia
Morocco
Mozambique
Myanmar

Namibia
Nepal
Netherlands
New Zealand
Nicaragua
Niger
Nigeria
Norway

Oman

Pakistan
Panama
Papua New Guinea
Paraguay
Peru
Philippines
Poland
Portugal

Qatar

Republic of Korea
Republic of Moldova
Romania
Russian Federation
Rwanda

Saint Kitts and Nevis
Saint Lucia
Saint Vincent and the Grenadines
Samoa
Saudi Arabia
Senegal
Seychelles
Sierra Leone
Singapore
Slovakia
Slovenia
Solomon Islands
South Africa
Spain
Sri Lanka
Sudan
Suriname
Swaziland
Sweden
Switzerland
Syrian Arab Republic

Thailand
The former Yugoslav Republic of
Macedonia
Togo
Tonga
Trinidad and Tobago
Tunisia
Turkey

Uganda
Ukraine
United Arab Emirates
United Kingdom
United Republic of Tanzania
United States of America
Uruguay

Vanuatu
Venezuela (Bolivarian Republic of)
Vietnam

Yemen

Zambia
Zimbabwe

The United States' position regarding food imports is that the country of origin must have a food safety management system, or HACCP plan, that is equal to or better than the requirements expected of a company in the United States. This raises the standards and expectations and gives countries the opportunity to sell their food products in the United States as long as these expectations are met. Some countries are distressed and look at this as a measure to impede imported products from entering certain countries.

International cooperation also helps in traceability if a contamination occurs. Various countries can work together to identify, control, eliminate, and then recover from the adulteration whether deliberate or accidental. **Traceability** is the ability to trace the history, application, or location of an item or activity with the help of documentation. Without using a precise tracking system, it would be difficult to track defective wheat from a loaf of bread back to the farm or the contaminated hamburger patty back to a specific cow.

Additional benefits of traceability are improved supply management initiatives with electronic coding systems, increased inventory control, and methodical tracking. Retail operators often require reputable suppliers to use third-party certification and documentation to verify their traceability system. If this tracking is in place, the supplier can isolate quality issues and, more importantly, the source and extent of any food safety problems. Following are some tracking methods to facilitate traceability from farm to table:

★ Providing country-of-origin labeling

★ Evaluating production process such as free-range or dolphin-safe

★ Requiring all manufacturers and processors to register and identify sources

These controls help to ensure the safety of the flow of food starting at the very source of the food chain. Without these screening tools, substandard (or adulterated or deleterious) foods will make their way to consumers. All controls and safeguards cannot be implemented at the last step in the flow of food—the selling or serving. Ensuring safe food requires a global team that implements controls and safeguards from farm to table.

To understand why HACCP is so important, you should familiarize yourself with its history. The Pillsbury Company, with the cooperation and participation of the National Aeronautic and Space Administration (NASA), Natick Laboratories of the U.S. Army, and the U.S. Air Force Space Laboratory Project Group, originally developed HACCP for the U.S. space program in the early 1960s. Yes, the HACCP program was actually developed by rocket scientists! They developed the program to prevent the astronauts from getting sick in space. Can you imagine an astronaut in space who is vomiting and has diarrhea? "Houston, we have a problem!"

Since all food contains microorganisms, NASA needed a food safety system that would prevent, eliminate, and reduce the number of microorganisms to safe levels to stop the astronauts from being stricken with a foodborne illness while in space. The United States space program approached 100 percent assurance against contamination by bacterial and viral pathogens, toxins, and chemical or physical hazards that could cause illness or injury to astronauts by implementing a successful HACCP plan. As a result, HACCP replaced end-product testing and provided a **preventive** system for producing safe food that had universal application.

Courtesy PhotoDisc, Inc.

After NASA incorporated HACCP plans, the military, manufacturers, schools, and, in some jurisdictions, retailers have also followed suit. In the succeeding years, the HACCP system has been recognized worldwide as an effective system of controls. The system has undergone considerable analysis, refinement, and testing and is widely accepted in the United States and internationally.

Established in 1988, the **National Advisory Committee on Microbiological Criteria for Foods (NACMCF)** is an advisory committee chartered under the **U.S. Department of Agriculture (USDA)** and composed of participants from the USDA (Food Safety and Inspection Service), Department of Health and Human Services (U.S. Food and Drug Administration and the Centers for Disease Control and Prevention), the Department of Commerce (National Marine Fisheries Service), the Department of Defense (Office of the Army Surgeon General), academia, industry, and state employees. NACMCF provides guidance and recommendations to the Secretary of Agriculture and the Secretary of Health and Human Services regarding the microbiological safety of foods.

In November 1992, NACMCF defined seven widely accepted HACCP principles that you should consider when developing a HACCP plan. In 1997, the NACMCF requested that the HACCP Working Group reconvene to review the committee's November 1992 HACCP document and compare it to current HACCP guidelines prepared by the Codex Alimentarius Committee on Food Hygiene. As a result of this 1997 committee meeting, HACCP was defined as a systematic approach to the identification, evaluation, and control of food safety hazards based on the following seven principles:

★ Principle 1: Conduct a hazard analysis.

★ Principle 2: Determine the critical control points (CCPs).

★ Principle 3: Establish critical limits.

★ Principle 4: Establish monitoring procedures.

★ Principle 5: Establish corrective actions.

★ Principle 6: Establish verification procedures.

★ Principle 7: Establish record-keeping and documentation procedures.

The 2005 edition of the FDA Model Food Code is the most recent edition created by the Food and Drug Administration (FDA) and the Centers for Disease Control and Prevention (CDC) of the U.S. Department of Health and Human Services (HHS) and the Food Safety and Inspection Service of the U.S. Department of Agriculture. The FDA Model Food Code is updated every 4 years. It is critical for all food safety management systems, especially your HACCP plan, to have the most current scientific information. The FDA Model Food Code provides practical, science-based guidance and manageable, enforceable provisions for mitigating risk factors known to cause foodborne illness. This Model Food Code is a reference document for retail food stores, other food establishments at the retail level, and institutions, such as schools, hospitals, nursing homes, child-care centers, and regulatory agencies that ensures food safety in foodservice establishments.

THE HACCP PHILOSOPHY

Now that you know the history of HACCP and why it is necessary to incorporate a HACCP plan, let's take a look at the HACCP philosophy. HACCP is internationally accepted. The Codex Alimentarius Code of Practice recommends a HACCP-based approach wherever possible to enhance food safety. It is important to recognize that HACCP is not a process conducted by an individual; it involves the entire **team**. This is why you are a part of this training session. Your company or institution is counting on you to do your part in preventing foodborne illness in your foodservice operation and in your part of the world.

It is critical that each facility assembles a HACCP team. Your designated HACCP team should not be larger than six members, including a team leader. However, some foodservice operations may need to enlarge the team temporarily with personnel from other departments such as purchasing, finance, marketing, and research and development. The HACCP team leader needs to assign responsibilities, train employees, and ensure the initiatives are accomplished. In addition, the leader needs to represent the HACCP team to management and request the financial and resource needs to enable a successful HACCP plan. These needs may include labor, time, materials, and money. The ideal candidates for a HACCP team should be skilled at identifying hazards, determining critical control points, monitoring, and verifying the prerequisite programs and the HACCP plan. HACCP team members should include people who have an understanding of the equipment used, aspects of the food operation, food microbiology, and HACCP principles. The first responsibility of the HACCP team should be to determine the scope of the HACCP plan. The team should

★ Define the flow of food to be studied.

★ Be specific with the processes and the products.

★ Identify the biological, chemical, and physical hazards to be included.

The **HACCP philosophy** simply states that biological, chemical, or physical hazards, at certain points in the flow of food, can either be

★ Prevented,

★ Removed, or

★ Reduced to safe levels.

Today, foods are transported around the world more than ever before. As a result, more people are handling the food products. The more food products are touched by people or machines, the greater the opportunity for contamination or, even worse, the greater the opportunity to spread a foodborne illness.

In the past, contaminated food only had the potential to affect the family that consumed it, or it could be spread throughout a village. Today, however, international trade and travel have the potential to quickly transport bacteria, viruses, or other microorganisms all over the world, creating not only epidemic but pandemic diseases. For example, consider bovine spongiform encephalopathy (BSE) or mad cow disease and avian influenza A (H5N1). When people eat BSE-infected animals, thus far presumed to be cows, they may develop the human version of mad cow disease, a variant called Creutzfeldt-Jakob disease (nvCJD). For this reason, millions of cattle suspected of being infected with BSE in England, Scotland, Ireland, France, Belgium, Italy, and other countries have been incinerated, and the United States, Europe, and Japan have instituted various safeguards to prevent a mad cow epidemic among humans as well as animals. Not surprisingly, the beef industry trade between the United States, Canada, and Japan has been considerably disrupted by the existence of mad cow disease.

According to the CDC, the current outbreak of avian influenza A (H5N1) or the bird flu that is spreading out from Southeast Asia is affecting poultry. Human consumption of avian-flu-infected poultry has caused infections and, in some cases, deaths. It is imperative that people avoid contact with avian-flu-infected birds or contaminated surfaces, and they should be careful when handling and cooking poultry as well. Many scientists believe it is only a matter of time until the next influenza pandemic occurs. The severity of the next pandemic cannot be predicted, but modeling studies suggest that its effect could be severe. World leaders, the medical community, shipping inspectors, and even the military are taking the threat of a pandemic seriously, and pressure is on to develop effective containment measures and treatments as preparation for a potential worldwide outbreak of the avian flu. Additional information and clarification of bovine spongiform encephalopathy, or mad cow disease, and the avian flu are provided by the CDC in the resource section of this book.

The eating habits of people around the world have also changed. People eat more ready-to-eat foods (RTE foods) and enjoy more ethnic dishes and food varieties than ever before, making it increasingly important for food establishments to apply HACCP plans to their food operations to ensure the safe consumption of their menu items. If more and more food establishments consistently apply HACCP plans, food will become safer for all consumers throughout the world. **This is why we need HACCP!**

The CDC lists the five most common risk factors that create foodborne illness:

★ Practicing poor personal hygiene

★ Improperly cooking food

★ Holding foods at the wrong temperatures

★ Using equipment that hasn't been properly calibrated, cleaned, and sanitized

★ Buying food from unsafe suppliers

Traditionally, food safety problems within establishments have not been addressed unless an observant, caring employee, a disgruntled customer, or an inspection by a regulatory agency has brought them to management's attention. Management has been reactive to food safety problems rather than striving to prevent these food safety problems during the day-to-day operation of the establishment. Preventing dangers is always better than having to react to the consequences. HACCP provides continual self-inspection, so regulatory agencies have ready access to documentation that food safety is practiced at all times, instead of relying on the specific moment of an on-site inspection that alerts management of out-of-control procedures.

Let's get started with Principle 1.

PRINCIPLE 1: CONDUCT A HAZARD ANALYSIS

A **hazard analysis** is the first principle in evaluating foods in your operation. The analysis looks for food safety hazards (biological, chemical, and physical) that are likely to occur in your operation if not effectively controlled. According to the 2005 FDA Food Code, a **hazard** is a biological, chemical, or physical property that may cause a food to be unsafe for human consumption. **Following are several questions that you may ask yourself when assessing the food safety of your operation:**

★ Does the food permit survival or multiplication of pathogens and/or toxin formation in the food before or during preparation?

★ Will the food permit survival or multiplication of pathogens and/or toxin formation during subsequent steps of preparation?

★ What has been the safety record for the product in the marketplace?

★ Is there an epidemiological history associated with this food?

★ Is the food served to a highly susceptible population?

★ What is known about the time/temperature exposure of the food?

★ What is the water activity and pH of the food?

★ Have bare hands touched the food, or otherwise cross-contaminated it?

★ Is the food from a safe source?

★ Do food workers practice good personal hygiene, including frequent and effective hand washing?

★ Has the food been exposed to unclean or unsanitized equipment?

★ Does the preparation procedure or process include a step that destroys pathogens or their toxins?

★ Is the product subject to recontamination after cooking?

By answering these questions, you are making an assessment of the hazards to determine the likelihood that a foodborne illness will occur. These foodborne illnesses are either foodborne infections, foodborne intoxications, or foodborne toxin-mediated infections. Let's look at each type of foodborne illness:

A **foodborne infection** is caused by ingesting **pathogens** or disease-causing microorganisms. When a person consumes enough of these pathogens, the person gets a foodborne infection as a result of their multiplication. Some examples of foodborne infections are salmonella, with the onset of symptoms occurring 8 to 72 hours following consumption, and listeriosis, which takes 3 to 21 days to affect someone after the unsafe food is consumed. Symptoms for a foodborne infection take longer to occur than foodborne intoxication because the disease-causing microorganisms need time to multiply.

Foodborne intoxication is caused by ingesting a **toxin** that is present in food. Toxins are poisons. When someone ingests a toxin, the first symptoms that occur very quickly are nausea and vomiting because the body is trying to eliminate the poison from the body as quickly as possible. This is actual "food poisoning." We tend to use food poisoning as a generic term to describe any foodborne illness because most people tend to assume that if they are sick, the most recent food they ingested caused it. In fact, most foodborne illnesses are foodborne infections, which generally have incubation ranges of 6 hours to 21 days.

Staphylococcus aureus, a foodborne intoxication, has a short incubation period of 1 to 6 hours with the duration of 24 to 48 hours. Besides nausea and vomiting, diarrhea and dehydration are common symptoms for Staphylococcus aureus. Foodborne intoxication is similar to alcohol intoxication because when a person consumes alcohol, the alcohol is a poison and affects the person immediately. If too much alcohol is consumed, then vomiting typically occurs, similar to foodborne intoxication. As a quick review, foodborne intoxication is caused by ingesting a toxin or a poison in food, the symptoms appear very quickly, and usually vomiting is one of the first symptoms to occur.

A **toxin-mediated infection** is an infection that is very severe because it is a combination of **pathogens and toxins**. The person ingests the pathogens, and then the infection produces toxins that begin to affect and/or damage organs in the body. An example of a toxin-mediated infection is hemorrhagic colitis, known as E. coli 0157:H7. The onset of this disease occurs 12 to 72 hours after food consumption, with symptoms of bloody diarrhea, severe abdominal cramping, and kidney failure. Toxins like verotoxin and shiga-like toxins cause severe damage to the lining of the intestines.

■ BIOLOGICAL HAZARDS

Biological hazards include bacterial, viral, and parasitic microorganisms. These microorganisms are of the most significant concern with regard to food safety. The CDC has identified more than 250 foodborne illnesses, where the majority are foodborne infections caused by bacteria, viruses, and parasites. The majority of biological hazards are **bacteria** that can be controlled through time, temperature, acidity, and water activity. Some bacteria form spores that are highly resistant and may not be destroyed by cooking and drying.

Viruses can exist in food without growing, but they can rapidly reproduce once they are on a living host, most typically a human being. Viruses can best be controlled by good personal hygiene, because that limits the transmission of viruses from one source (human) to another via human contact or common food contact. Norovirus, hepatitis A, and rotavirus are directly related to contamination from human feces. This sort of contamination can be prevented by forbidding bare-hand contact with food, requiring proper hand washing, and avoiding contact with ill employees. Once the virus has a host, the cells grow and reproduce, and cause illness.

Parasites also need a host. They are mostly animal-host-specific, but they can survive in humans. Adequate cooking or freezing destroys parasites. So special attention should be given when preparing foods such as pork, fish, and bear, since they are known to carry parasites.

Annex 4, Table 1. Selected Biological Hazards Found at Retail, Associated Foods, and Control Measures		
Hazard	**Associated Foods**	**Control Measures**
Bacteria		
Bacillus cereus (intoxication caused by heat stable, preformed emetic toxin and infection by heat labile, diarrheal toxin)	Meat, poultry, starchy foods (rice, potatoes), puddings, soups, cooked vegetables	Cooking, cooling, cold holding, hot holding
Campylobacter jejuni	Poultry, raw milk	Cooking, handwashing, prevention of cross-contamination
Clostridium botulinum	Vacuum-packed foods, reduced oxygen packaged foods, under-processed canned foods, garlic-in-oil mixtures, time/temperature abused baked potatoes/sautéed onions	Thermal processing (time + pressure), cooling, cold holding, hot holding, acidification and drying, etc.
Clostridium perfringens	Cooked meat and poultry, Cooked meat and poultry products including casseroles, gravies	Cooling, cold holding, reheating, hot holding
E. coli 0157:H7 (other shiga toxin-producing *E. coli*)	Raw ground beef, raw seed sprouts, raw milk, unpasteurized juice, foods contaminated by infected food workers via fecal-oral route	Cooking, no bare hand contact with RTE foods, employee health policy, handwashing, prevention of cross-contamination, pasteurization or treatment of juice
Listeria monocytogenes	Raw meat and poultry, fresh soft cheese, paté, smoked seafood, deli meats, deli salads	Cooking, date marking, cold holding, handwashing, prevention of cross-contamination
Salmonella spp.	Meat and poultry, seafood, eggs, raw seed sprouts, raw vegetables, raw milk, unpasteurized juice	Cooking, use of pasteurized eggs, employee health policy, no bare hand contact with RTE fods, handwashing, pasteurization or treatment of juice
Shigella spp.	Raw vegetables and herbs, other foods contaminated by infected workers via fecal-oral route	Cooking, no bare hand contact with RTE foods, employee health policy, handwashing

(continues)

Annex 4, Table 1. Continued

Hazard	Associated Foods	Control Measures
Staphylococcus aureus (preformed heat stable toxin)	RTE PHF foods touched by bare hands after cooking and further time/temperature abused	Cooling, cold holding, hot holding, no bare hand contact with RTE food, handwashing
Vibrio spp.	Seafood, shellfish	Cooking, approved source, prevention of cross-contamination, cold holding
Parasites		
Anisakis simplex	Various fish (cod, haddock, fluke, pacific salmon, herring, flounder, monkfish)	Cooking, freezing
Taenia spp.	Beef and pork	Cooking
Trichinella spiralis	Pork, bear, and seal meat	Cooking
Viruses		
Hepatitis A and E	Shellfish, any food contaminated by infected worker via fecal-oral route	Approved source, no bare hand contact with RTE food, minimizing bare hand contact with foods not RTE, employee health policy, handwashing
Other Viruses (Rotavirus, Norovirus, Reovirus)	Any food contaminated by infected worker via fecal-oral route	No bare hand contact with RTE food, minimizing bare hand contact with foods not RTE, employee health policy, handwashing

RTE = ready- to-eat
PHF = potentially hazardous food (time/temperature control for safety food)
Source: 2005 FDA code http://www.cfsan.fda.gov/~acrobat/fc05-a4.pdf, page 481.

■ CHEMICAL HAZARDS

Chemical hazards may also cause foodborne illness. Chemical hazards may occur naturally or may be introduced during any stage of food production. Dangerous, naturally occurring chemicals can be found in some species of fish (scombroid, ciguatera, puffer fish) or shellfish (molluscan, lobsters, red rock crabs), some plant foods (red kidney beans), or mushrooms, and allergens. Allergens can be biological or chemical hazards and were previously discussed in Star Point 1.

Some chemicals added to foods also make them unsafe. These include preservatives (sulfites, sodium nitrates, monosodium glutamate, or MSG), environmental additives (fertilizers, pesticides), and cleaning agents (sanitizers, lubricants). High levels of toxic chemicals may cause acute cases of foodborne illness, while chronic illness may result from lower levels.

Annex 4, Table 2. Common Chemical Hazards at Retail, Along with Their Associated Foods and Control Measures		
Chemical Hazards	**Associated Foods**	**Control measures**
Naturally Occurring:		
Scombrotoxin	Primarily associated with tuna fish, mahi-mahi, blue fish, anchovies bonito, mackerel; Also found in cheese	Check temperatures at receiving; store at proper cold holding temperatures; buying specifications: obtain verification from supplier that product has not been temperature abused prior to arrival in facility.
Ciguatoxin	Reef fin fish from extreme SE US Hawaii, and tropical areas; barracuda, amberjacks, king mackerel, large groupers, and snappers	Ensure fin fish have not been caught: • Purchase fish from approved sources. • Fish should not be harvested from an area that is subject to an adverse advisory.
Tetrodoxin	Puffer fish (Fugu; Blowfish)	Do not consume these fish.
Mycotoxins Aflatoxin	Corn and corn products, peanuts and peanut products, cottonseed, milk, and tree nuts such as Brazil nuts, pecans, pistachio nuts, and walnuts. Other grains and nuts are susceptible but less prone to contamination.	Check condition at receiving; do not use moldy or decomposed food.
Patulin	Apple juice products	Buyer Specification: obtain verification from supplier or avoid the use of rotten apples in juice manufacturing.
Toxic mushroom species	Numerous varieties of wild mushrooms	Do not eat unknown varieties or mushrooms from unapproved source.
Shellfish toxins Paralytic shellfish poisoning (PSP)	Molluscan shellfish from NE and NW coastal regions; mackerel, viscera of lobsters and Dungeness, tanner, and red rock crabs	Ensure molluscan shellfish are: • from an approved source; and • properly tagged and labeled.
Diarrhetic shellfish poisoning (DSP)	Molluscan shellfish in Japan, western Europe, Chile, NZ, eastern Canada	
Neurotoxin shellfish poisoning (NSP)	Molluscan shellfish from Gulf of Mexico	
Amnesic shellfish poisoning (ASP)	Molluscan shellfish from NE and NW coasts of NA; viscera of Dungeness, tanner, red rock crabs and anchovies.	
Pyrrolizidine alkaloids	Plants food containing these alkaloids. Most commonly found in members of the Borginaceae, Compositae, and Leguminosae families.	Do not consume of food or medicinals contaminated with these alkaloids.

Annex 4, Table 2. Continued		
Chemical Hazards	**Associated Foods**	**Control measures**
Naturally Occurring (cont.):		
Phtyohaemmagglutinin	Raw red kidney beans (Undercooked beans may be more toxic than raw beans)	Soak in water for at least 5 hours. Pour away the water. Boil briskly in fresh water, with occasional stirring, for at least 10 minutes.
Added Chemicals:		
Environmental contaminants: Pesticides, fungicides, fertilizers, insecticides, antibiotics, growth hormones	Any food may become contaminated.	Follow label instructions for use of environmental chemicals. Soil or water analysis may be used to verify safety.
PCBs	Fish	Comply with fish advisories.
Prohibited substances (21 CFR 189)	Numerous substances are prohibited from use in human food; no substance may be used in human food unless it meets all applicable requirements of the FD&C Act.	Do not use chemical substances that are not approved for use in human food.
Toxic elements/compounds Mercury	Fish exposed to organic mercury: shark, tilefish, king mackerel and swordfish. Grains treated with mercury based fungicides	Pregnant women/women of childbearing age/nursing mothers, and young children should not eat shark, swordfish, king mackerel or tilefish because they contain high levels or mercury. Do not use mercury containing fungicides on grains or animals.
Copper	High acid foods and beverages.	Do not store high acid foods in copper utensils; use backflow prevention device on beverage vending machines.
Lead	High acid food and beverages.	Do not use vessels containing lead.
Preservatives and Food Additives: Sulfiting agents (sulfur dioxide, sodium and potassium bisulfite, sodium and potassium metabisulfite)	Fresh fruits and Vegetables Shrimp Lobster Wine	Sulfiting agents added to a product in a processing plant must be declared on labeling. Do not use on raw produce in food establishments.
Naturally Occurring:		
Nitrites/nitrates Niacin	Cured meats, fish, any food exposed to accidental contamination, spinach Meat and other foods to which sodium nicotinate is added	Do not use more than the prescribed amount of curing compound according to labeling instructions. Sodium nicotinate (niacin) is not currently approved for use in meat or poultry with or without nitrates or nitrates.
Flavor enhancers Monosodium glutamate (MSG)	Asian or Latin American food	Avoid using excessive amounts

(continues)

Annex 4, Table 2. Continued		
Chemical Hazards	**Associated Foods**	**Control measures**
Naturally Occurring (cont.):		
Chemicals used in retail establishments (e.g., lubricants, cleaners, sanitizers, cleaning compounds, and paints	Any food could become contaminated	Address through SOPs for proper labeling, storage, handling, and use of chemicals; retain Material Safety Data Sheets for all chemicals.
Allergens	Foods containing or contacted by: Milk Egg Fish Crustacean shellfish Tree nuts Wheat Peanuts Soybeans	Use a rigorous sanitation regime to prevent cross contact between allergenic and non-allergenic ingredients.

■ PHYSICAL HAZARDS

A physical hazard is any physical material or foreign object not normally found in a food that can cause illness and injury. These physical hazards may be the result of contamination, carelessness, mishandling, or implementing poor procedures at many points in the food chain from harvest to consumer, including those within the food establishment. These hazards are the easiest to identify because the consumer usually finds the foreign object and reports the incident.

Annex 4, Table 3. Main Materials of Concern as Physical Hazards and Common Sources[a, b]		
Material	**Injury Potential**	**Sources**
Glass fixtures	Cuts, bleeding; may require surgery to find or remove	Bottles, jars, light, utensils, gauge covers
Wood	Cuts, infection, choking; may require surgery to remove	Fields, pallets, boxes, buildings
Stones, metal fragments	Choking, broken teeth Cuts, infection; may require surgery to remove	Fields, buildings, machinery, wire, employees
Insulation	Choking; long-term if asbestos	Building materials
Bone	Choking, trauma	Fields, improper plant processing
Plastic	Choking, cuts, infection; may require surgery to remove	Fields, plant packaging materials, pallets, employees
Personal effects	Choking, cuts, broken teeth; may require surgery to remove	Employees

[a] Adapted from Corlett (1991).
[b] Used with permission, "HACCP Principles and Applications," Pierson and Corlett, Eds. 1992. Chapman & Hall, New York, NY.

■ HAZARD ANALYSIS: A TWO-STAGE PROCESS

It is important to always remember that **hazard analysis** focuses on food safety, not food quality. It is equally important to separate food safety from food quality. Consider the following: One or more reheating steps might dry out a certain food. Can this food still be served? By HACCP standards, the answer is yes. While it may have lost some of its appeal to a customer, it is still safe to eat. The process of conducting a hazard analysis involves two stages. Stage 1 is **hazard identification** and stage 2 is **hazard evaluation** of the foods that are used in your establishment. Managers should focus on the two stages in the hazard analysis:

Stage 1: Hazard Identification

A. **Intended use.** Analyze your customers and the consumers of the product.

B. **Menu.** Evaluate your menu, including the facility and equipment used.

C. **General characteristics of the food.** Identify ingredients used and potentially hazardous food (time/temperature control for safety of food).

D. **Preparation process and activities conducted at each step.** Understand the flow of food to determine where hazards may be controlled.

E. **Handling the food.** Divide your menu items into three categories by how the food is prepared: simple/no-cook recipes, same-day recipes, and complex recipes.

Stage 2: Hazard Evaluation

F. **Analyze two questions:**

"What is the likelihood of a hazard to occur here?"

"What is the risk if the hazard does occur?"

Overview of Stage 1: Hazard Identification

This stage focuses on identifying the food safety hazards that might be present in the food given the food preparation process used, the handling of the food, the facility, and general characteristics of the food itself. During this stage, a review of the ingredients used in the product, the activities conducted at each step in the process, the equipment used, the final product, and its method of storage and distribution, as well as the intended use and consumers of the product is made. Based on this review, a list of potential biological, chemical, or physical hazards is made at each stage in the food preparation process.

Overview of Stage 2: Hazard Evaluation

In Stage 2, the hazard evaluation, each potential hazard is evaluated based on the severity of the potential hazard and its likely occurrence. The purpose of this stage is to determine which of the potential hazards listed in Stage 1 of the hazard analysis warrant control in the HACCP plan. As mentioned, answers to two questions help to determine the status of any hazard:

★ "What is the likelihood of a hazard to occur here?"

★ "What is the risk if the hazard does occur?"

Severity is the seriousness of the consequences if exposure to the hazard occurs. When determining the severity of a hazard, you must recognize the impact of the medical condition caused by the illness, as well as the magnitude and duration of the illness or injury. Consideration of the likely occurrence is usually based upon a combination of experience, epidemiological data, and information in the technical literature. During the evaluation of each potential hazard, the food; its method of preparation, transportation, storage; and persons likely to consume the product should be considered to determine how each of these factors may influence the likely occurrence and severity of the hazard being controlled. For example, E. coli 0157:H7 is definitely a potential hazard in raw ground beef; it is likely to occur. However, when properly cooked to 155°F (68.3°C) for 15 seconds, the likelihood of E. coli 0157:H7 presenting a severe hazard is reduced. Also, the presence of E. coli 0157:H7 in precooked frozen hamburger patties has already been reduced, so the likelihood of a severe hazard is not present for E. coli 0157:H7. The patty that is commercially processed, hermetically sealed, or provided in an intact package from a food processing plant must still be reheated to 135°F (57.2°C) for 15 seconds to reduce other microorganisms that may have been introduced to the food via food handlers, preparation, or packaging.

Here is further information on completing these two stages.

More on Stage 1: Hazard Identification

A. **Intended use: Analyze your customers and the consumers of the product.**

Do you serve the general population? Or do you serve a highly susceptible population? Schools, hospitals, day-care sites for children and adults, and assisted-living centers all require very strict standards of operation because of the vulnerability of these customers. You must be aware of groups like infants and children, pregnant women, people on medications, or an aged population when conducting hazard identification.

STAR KNOWLEDGE EXERCISE: ANALYZE CUSTOMERS

Mark an "X" in the box that analyzes the customer bases as examples of general population or a highly susceptible population:

Customer Base	General Population	Highly Susceptible Population
★ Convenience store		
★ Middle school		
★ Quick-service restaurant		
★ Senior-care facility		
★ Hospital		
★ Casual-dining restaurant		
★ Fine-dining restaurant		
★ Day care		
★ Institutional foodservice		
★ Tavern		

B. Evaluate your menu including the facility and equipment that is used.

Compare your menu to the ability of your employees to maintain a HACCP plan. Do you have employees who are capable of handling complex menu items and procedures? Or do you need to train or hire people with more advanced culinary training? Are your kitchen and equipment capable of producing the foods you want on your menu? In evaluating these situations, you may need to refine, reduce, or simplify your menu. It is advantageous to address such questions now, as repurchasing, redesigning, and rehiring are expensive alternatives to a failed plan.

STAR KNOWLEDGE EXERCISE: EVALUATE YOUR MENU

Using the sample establishments in the previous activity, determine if the menu items, equipment, facility, and employee qualifications are appropriate. Mark an "X" in the box for acceptable, unacceptable, and needs improvement.

Convenience Store	Acceptable	Unacceptable	Needs Improvement
Typical menu items:			
Equipment needed:			
Facility size:			
Employee qualifications:			
Middle School	Acceptable	Unacceptable	Needs Improvement
Typical menu items:			
Equipment needed:			
Facility size:			
Employee qualifications:			

Quick-Service Restaurant	Acceptable	Unacceptable	Needs Improvement
Typical menu items:			
Equipment needed:			
Facility size:			
Employee qualifications:			

Senior-Care Facility	Acceptable	Unacceptable	Needs Improvement
Typical menu items:			
Equipment needed:			
Facility size:			
Employee qualifications:			

Hospital	Acceptable	Unacceptable	Needs Improvement
Typical menu items:			
Equipment needed:			
Facility size:			
Employee qualifications:			

Casual-Dining Restaurant	Acceptable	Unacceptable	Needs Improvement
Typical menu items:			
Equipment needed:			
Facility size:			
Employee qualifications:			

Fine-Dining Restaurant	Acceptable	Unacceptable	Needs Improvement
Typical menu items:			
Equipment needed:			
Facility size:			
Employee qualifications:			

Day Care	Acceptable	Unacceptable	Needs Improvement
Typical menu items:			
Equipment needed:			
Facility size:			
Employee qualifications:			

Institutional Food Service	Acceptable	Unacceptable	Needs Improvement
Typical menu items:			
Equipment needed:			
Facility size:			
Employee qualifications:			
Tavern	Acceptable	Unacceptable	Needs Improvement
Typical menu items:			
Equipment needed:			
Facility size:			
Employee qualifications:			
Your Operation	Acceptable	Unacceptable	Needs Improvement
Typical menu items:			
Equipment needed:			
Facility size:			
Employee qualifications:			

C. **General characteristics of the food: Ingredients used and identification of potentially hazardous food (time/temperature control for safety of food).**

The first thing to do when analyzing the food in your establishment is to identify all potentially hazardous food (**PHF/TCS**). Using this sample menu and the recipes provided, can you identify any potentially hazardous foods (time/temperature control for safety food)? Remember, not all foods are potentially hazardous or need time/temperature control for safety.

SAMPLE MENU

Starters:

Chili
Tuna Spread

Sandwiches:

Pork Barbeque Sandwich
Tuna Melt

Side Dishes:

Spanish Rice Bake
Chicken Pasta Primavera

Dessert:

Mixed-Fruit Crisp
Banana Smoothie

STAR KNOWLEDGE EXERCISE: PHF/TCS

These are the recipes for the items in the sample menu above. List any PHFs /TCS from each recipe.

MIXED-FRUIT CRISP

1 15-ounce can (443.6 ml) mixed fruit

1/2 cup (118.29 ml) quick rolled oats

1/2 cup (118.29 ml) brown sugar

1/2 cup (118.29 ml) all-purpose flour

1/4 teaspoon (1.24 ml) baking powder

1/2 teaspoon (2.45 ml) ground cinnamon

1/4 cup (59.15 ml) butter or margarine

(Recommendation: prepare a day in advance.)

1. Preheat oven to 350°F (176.6°C).

2. Drain mixed fruit and set aside.

3. Lightly grease an 8- or 9-inch (20.32- or 22.86-cm) baking pan. Place the mixed fruit on the bottom of the pan.

4. In a smaller bowl, combine all of the dry ingredients. Cut in the butter or margarine with a pastry blender. Sprinkle mixture over mixed-fruit filling.

5. Bake for 30 to 35 minutes in conventional oven to a minimum internal temperature of 135°F (57.2°C) for 15 seconds.

6. Cool properly. Cool hot food from 135°F to 70°F (57.2°C to 21.1°C) within 2 hours; you then have an additional 4 hours to go from 70°F to 41°F (21.1°C to 5°C) or lower for a maximum total cool time of 6 hours.

7. Store in refrigeration at 41°F (5°C) or lower.

8. Reheat 165°F (73.9°C) for 15 seconds within 2 hours, serve warm.

PHF/TCS:

_____

_____

QUICK TUNA SPREAD

1 12-ounce can (354.88 ml) tuna in water, drained and flaked

1/3 cup (78.86 ml) sandwich spread (or 3 tablespoons sweet pickle relish (44.36 ml) and 1/3 cup (78.86 ml) mayonnaise)

1. In a small bowl, mix tuna and sandwich spread.

2. Store in refrigeration at 41°F (5°C) or lower.

Serving Ideas: Quick Tuna Spread can be served in many different ways:

- As a sandwich using whole wheat bread
- As a dip using pita bread cut into triangles
- As a snack rolled up in a flour tortilla

PHF/TCS:

_____

_____

BANANA SMOOTHIE

1 cup (236.59 ml) evaporated milk

1 ripe banana

1 teaspoon (4.91 ml) lemon juice

2 cups (473.18 ml) ice

1 tablespoon (14.79 ml) honey or sugar

optional nutmeg

1. Mix evaporated milk, banana, lemon juice, and honey together in a blender on high speed. Add ice gradually; process until slushy. Sprinkle with nutmeg, if desired.

PHF/TCS:

_____

_____

CHILI

12 ounces (340.19 grams) ground beef

1 cup (236.59 ml) onion, chopped (1 large onion)

1/2 cup (118.29 ml) green bell pepper, chopped

2 cloves garlic, minced

1 15.5-ounce can (458.39 ml) tomatoes, cut up (do not drain the liquid from the can)

1 15.5-ounce can (458.39 ml) dark red kidney beans, rinsed and drained

1 8-ounce can (236.59 ml) tomato sauce

2 to 3 teaspoons (9.82–14.73 ml) chili powder

1/2 teaspoon (2.45 ml) dried basil, crushed

1/4 teaspoon (1.24 ml) pepper

1. In a large saucepan, cook ground beef, onion, bell pepper, and garlic until meat is brown and onion is tender. Drain fat.

2. Stir in undrained tomatoes, kidney beans, tomato sauce, chili powder, basil, and pepper.

3. Cook to minimum internal temperature of 165°F (73.9°C) for 15 seconds.

4. Hold on stove for 1 hour.

5. Cool properly. Cool hot food from 135°F to 70°F (57.2°C to 21.1°C) within 2 hours; you then have an additional 4 hours to go from 70°F to 41°F (21.1°C to 5°C) or lower for a maximum total cool time of 6 hours.

6. Allow chili flavors to blend in refrigeration. Store in refrigeration at 41°F (5°C) or lower.

7. Properly reheat the next day to 165°F (73.9°C) for 15 seconds within 2 hours, serve.

PHF/TCS:

_____

_____

SPANISH RICE BAKE

1 pound (453.59 grams) lean ground beef

1/2 cup (118.29 ml) onion, finely chopped

1/4 cup (59.15 ml) green bell pepper, chopped

1 15.5-ounce can (458.39 ml) tomatoes

1 cup (236.59 ml) water

3/4 cup (177.44 ml) uncooked long-grain rice

1/4 cup (118.29 ml) chili sauce

1 teaspoon (4.91 ml) salt

1/2 teaspoon (2.45 ml) Worcestershire sauce

1 pinch ground black pepper

1/2 cup (118.29 ml) shredded cheddar cheese

1/2 teaspoon (2.45 ml) ground cumin (optional)

2 tablespoons (29.57 ml) chopped fresh cilantro (optional)

1. Preheat oven to 375°F (190.55°C).

2. Brown the ground beef in a large skillet over medium-high heat. Drain excess fat and transfer beef to a large pot over medium-low heat. Stir in the onion, green bell pepper, tomatoes, water, rice, chili sauce, salt, brown sugar, cumin, Worcestershire sauce, and ground black pepper.

3. Simmer for about 30 minutes, stirring occasionally, and then put into a 2-quart (1.892-liter) casserole dish. Press down firmly and sprinkle with the shredded cheddar cheese.

4. Bake to a minimum internal temperature of 165°F (73.9°C) for 15 seconds. Garnish with chopped fresh cilantro, if desired.

PHF/TCS:

_____

_____

CHICKEN AND PASTA PRIMAVERA

1-1/2 cups (354.88 ml) uncooked bow tie pasta (or any other type of pasta)

1 10.75-ounce can (317.92 ml) condensed cream of mushroom soup

3/4 cup (177.44 ml) milk

1/8 teaspoon (.62 ml) ground black pepper

2 cups (473.18 ml) broccoli florets

1/8 teaspoon (.67 ml) garlic powder

2 carrots, sliced thinly

1/3 can (about 10 ounces) (295.74 ml) canned chicken, drained

1/4 cup (59.15 ml) grated parmesan cheese (optional)

1. Cook pasta in boiling water. Drain.

2. While the pasta is cooking, prepare the cream sauce. In a medium saucepan, stir together soup, milk, pepper, broccoli, garlic powder, carrots, and parmesan cheese (optional). Reduce heat to low and cover. Simmer for 10 minutes, or until vegetables are tender. Stir occasionally.

3. Stir pasta and chicken into cream sauce and heat thoroughly to a minimum internal temperature of 165°F (73.9°C) for 15 seconds.

PHF/TCS:

_____

_____

10-MINUTE PORK BBQ SANDWICH

1 teaspoon (4.91 ml) vegetable oil

1 large onion, chopped

2 cups (473.18 ml) canned pork

3/4 cup (177.44 ml) prepared barbecue sauce

5 hamburger rolls

1. In large skillet, heat the oil on low heat.

2. Add onion and cook until tender, about 5 minutes.

3. Mix in pork and barbecue sauce and cook to a minimum internal temperature of 135°F (57.2°C) for 15 seconds.

4. Spoon barbecue mixture on bottom half of opened hamburger bun.

PHF/TCS:

_____

_____

TUNA MELT SANDWICH

1 12-ounce can (354.88 ml) tuna, drained and flaked

⅓ cup (78.86 ml) low-fat mayonnaise/mayonnaise

¼ teaspoon (1.24 ml) dry mustard

3 tablespoons (44.36 ml) minced fresh onion

½ cup (118.29 ml) finely diced celery

⅓ cup (5 ounces) (78.86 ml) shredded American cheese

5 English muffins, split

1. In a bowl, combine the dry mustard and mayonnaise.
2. Stir in the onions, celery, and drained tuna. Toss lightly to mix.
3. Mix in half of the shredded cheese.
4. Lay out split English muffins onto a baking pan. Spread ¼ cup (59.15 ml) of tuna salad to the edge of each muffin.
5. Sprinkle the top with 1 tablespoon (14.79 ml) of remaining shredded cheese.
6. Bake at 350°F (176.6°C) for 5 minutes until cheese is melted.

PHF/TCS:

_____

_____

D. Preparation process activities conducted at each step: Understand the flow of food to determine where hazards may be controlled.

In order to determine where hazards may be controlled, you need to understand the **flow of food**. All the food we eat goes through what we call a flow of food. All flows of food start with the purchasing of food from approved sources. The food is then received, stored, prepared, cooked, held, cooled, reheated, and served.

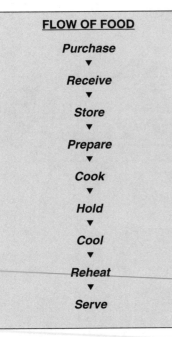

FLOW OF FOOD

Purchase
▼
Receive
▼
Store
▼
Prepare
▼
Cook
▼
Hold
▼
Cool
▼
Reheat
▼
Serve

The flow of food is best described by looking at how you make a pot of homemade soup. First, you **purchase** the ingredients from a clean, safe grocery store. You take the food you purchased and **receive** the food into your home; then you need to **store** the food properly either in dry storage (your cabinets or pantry) or cold storage (your refrigerator or freezer). Once the food is stored properly, you then **prepare** the food (slicing vegetables, portioning meats, etc.). After preparation, the product is cooked. To **cook** your homemade soup properly, you must reach a minimum internal cooking temperature of 165°F (73.9°C) for 15 seconds. Once the soup has reached 165°F (73.9°C), then you can **hold** the soup. The correct hot-holding temperature for your delicious homemade soup is 135°F (57.2°C).

The next step in the flow of food is to **cool** the soup for storage in your refrigerator. The proper way to cool soup is to cool it as quickly as possible to 41°F (5°C) (follow the cooling process you learned in Star Point 1). The next day, you see the soup in your refrigerator and you decide to **reheat** it. The soup must be reheated to 165°F (73.9°C) for 15 seconds within 2 hours to be safe. The final step is to **serve** the soup and enjoy.

In summary, the flow of food is purchase, receive, store, prepare, cook, hold, cool, reheat, and serve. HACCP helps you ensure that at every step in the flow of food, the food stays safe. In our example, no one got sick by eating the homemade soup because food safety SOPs were followed in the flow of food. It is necessary for all food establishments to set the same goal—to serve safe food.

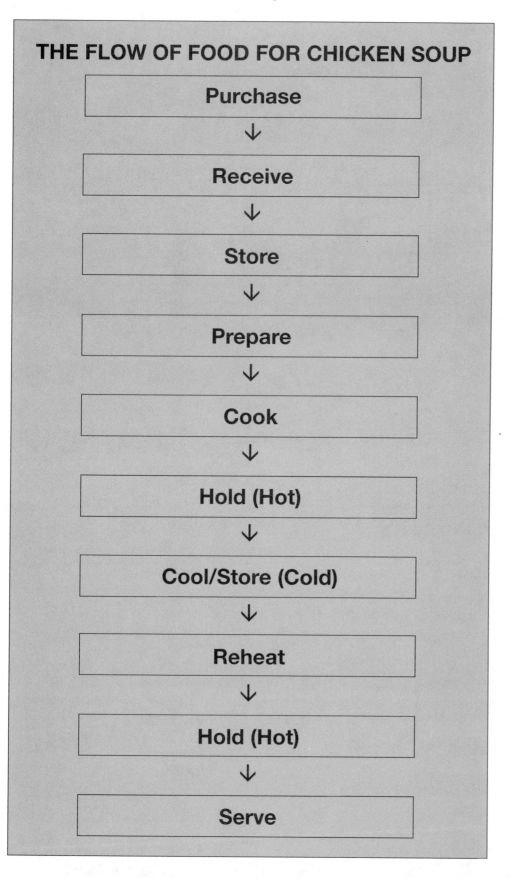

THE FLOW OF FOOD FOR CHICKEN SOUP

Purchase

↓

Receive

↓

Store

↓

Prepare

↓

Cook

↓

Hold (Hot)

↓

Cool/Store (Cold)

↓

Reheat

↓

Hold (Hot)

↓

Serve

STAR KNOWLEDGE EXERCISE: FLOW-OF-FOOD ACTIVITY

In this activity, identify the steps in the flow of food for each of the items in the sample menu. Some items will have more steps than others.

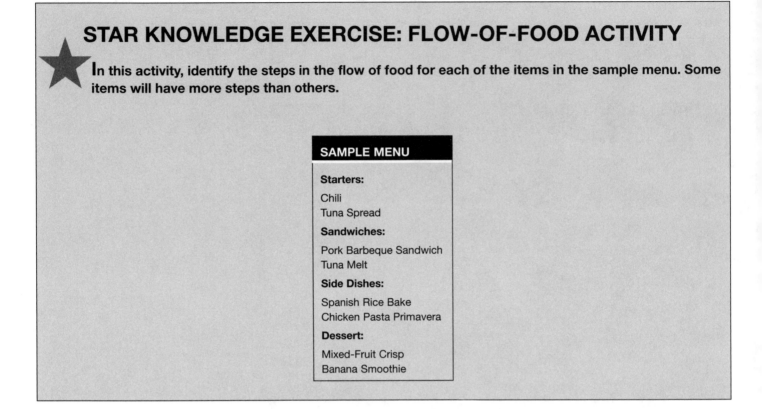

SAMPLE MENU

Starters:

Chili
Tuna Spread

Sandwiches:

Pork Barbeque Sandwich
Tuna Melt

Side Dishes:

Spanish Rice Bake
Chicken Pasta Primavera

Dessert:

Mixed-Fruit Crisp
Banana Smoothie

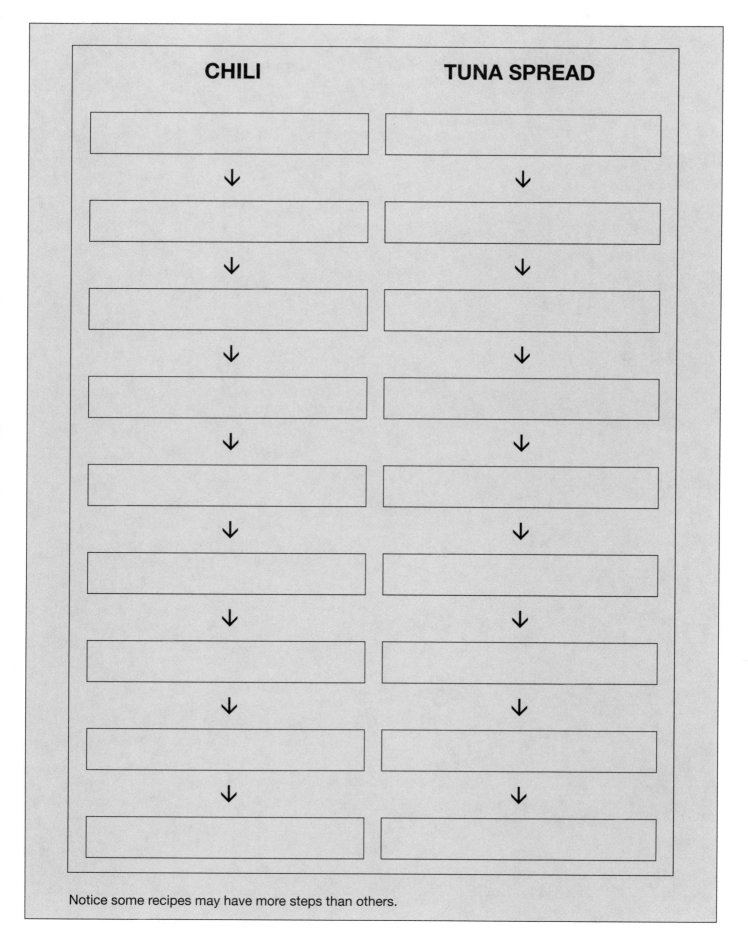

CHILI   **TUNA SPREAD**

Notice some recipes may have more steps than others.

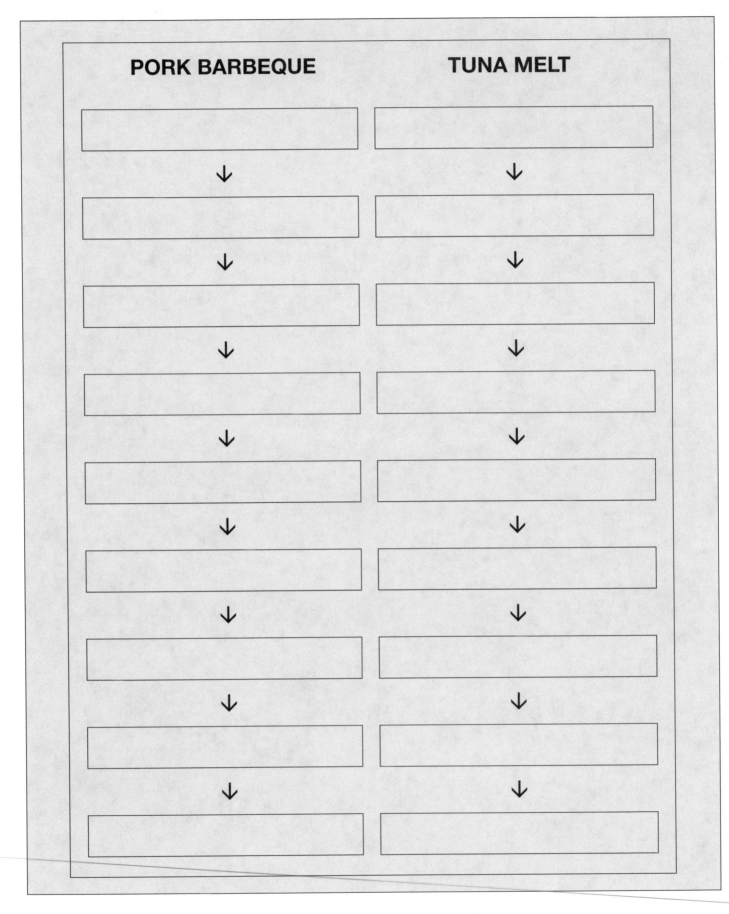

PORK BARBEQUE   **TUNA MELT**

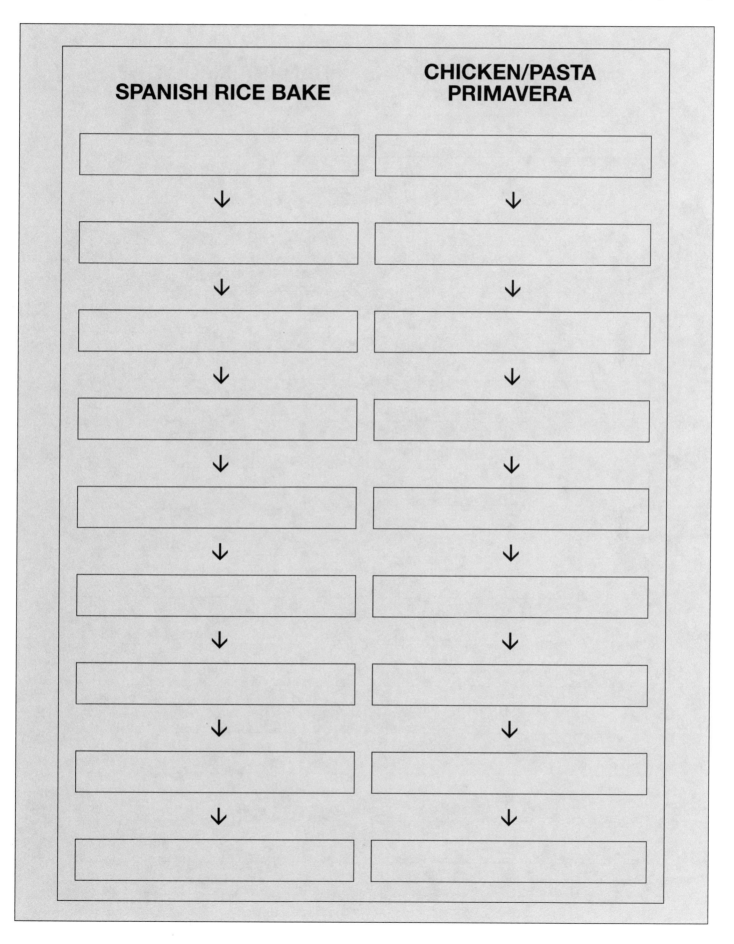

SPANISH RICE BAKE

CHICKEN/PASTA PRIMAVERA

MIXED-FRUIT CRISP

BANANA SMOOTHIE

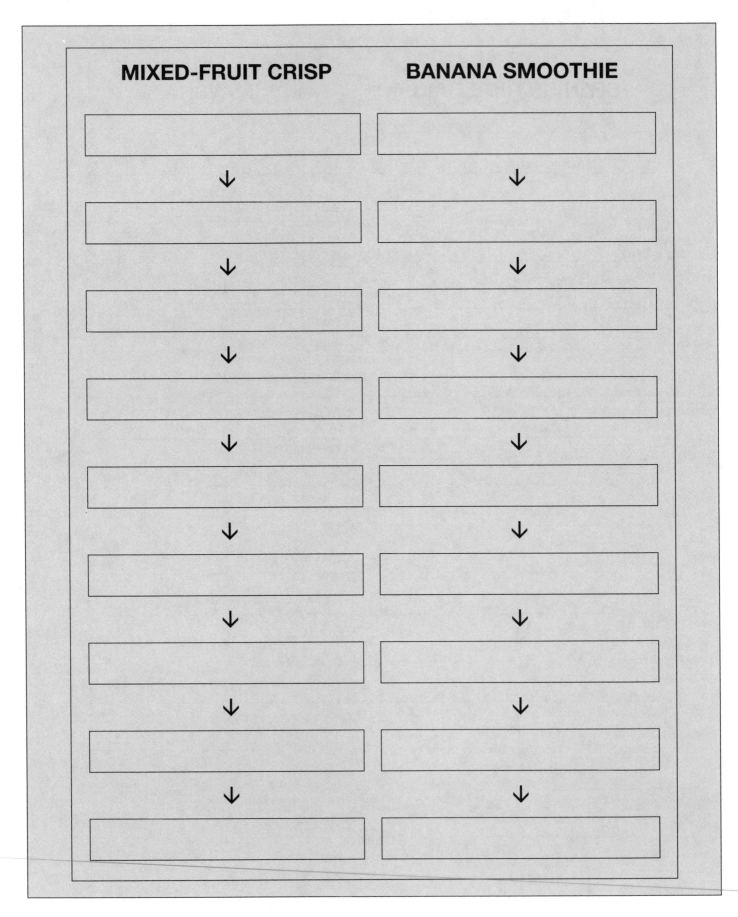

E. Handling the food: Divide your menu items into three categories by how the food is prepared: simple/no-cook recipes, same-day recipes, and complex recipes.

Once the flow of food is determined for your menu items, you should then divide your menu items into one of three categories by how the item is prepared and the number of times the food moves through the TDZ. There are three ways to divide the menu: no-cook, same-day, and complex.

A **no-cook/simple recipe** means exactly that: There is no cooking involved. For example, when tuna salad is prepared, a can or bag of tuna is opened, drained of the juice, placed in a bowl, and mayonnaise and seasonings are added. It is then mixed well, chilled, and served. These foods make no complete trips through the TDZ.

Other foods your HACCP plan might identify as no-cook/simple recipes include the following:

★ Deli meats   ★ Salads   ★ Cheese   ★ Raw oysters
★ Cole slaw   ★ Sashimi   ★ Yogurt   ★ Fresh fruit
★ Vegetable tray   ★ Retail sales of raw meat, steaks, poultry, and so on

Since there is no cook step to reduce harmful microorganisms, good hygiene is a very important control measure. **Control measures** are activities and actions that are used by foodservice and retail operators to prevent, eliminate, or reduce food safety hazards to an acceptable level.

You will need to determine the control measures that should be implemented to prevent the occurrence of risk factors in each food preparation process. Additional control measures are preventing cross-contamination, cooking, hot holding, cold holding, cooling, reheating, drying, pasteurization, and acidification.

A **same-day recipe** means a food product is prepared for same-day service or has some same-day cooking involved. The food will pass through the TDZ **once** before it is served or sold. For example, a hamburger requires that you take the frozen raw hamburger patty from the freezer, place the hamburger on the grill, cook the hamburger to 155°F (68.3°C) for 15 seconds, place the cooked hamburger on a bun, and serve.

Other foods your HACCP plan might identify as same-day recipes include the following:

★ Hamburger   ★ Baked chicken   ★ Baked meatloaf
★ Chicken wrap   ★ Fried chicken   ★ Cheesesteak
★ Popcorn shrimp   ★ Grilled cheese   ★ Scrambled eggs
★ Grilled vegetables ★ Beef fajita   ★ Shrimp stir-fry

A **complex recipe** calls for a food to be prepared, cooled, stored, and then reheated. If a food moves through the temperature danger zone more than **two times** it is considered a complex recipe. The homemade soup described earlier is an example of a complex recipe. When using a complex recipe you must

1. Determine the potential for microorganisms to
   a. Survive a heat process (cooking or reheating) and
   b. Multiply at room temperature and during hot and cold holding.

2. Find sources and specific points of contamination that are not covered in the food safety and food defense standard operating procedures. Other foods your HACCP plan might identify as complex recipes include the following:

★ Potato salad   ★ Lasagna   ★ Twice-baked potatoes

★ Chili   ★ Casserole   ★ Homemade meatballs

★ Chicken Armando (chicken stuffed with grilled eggplant)

As a summary, the 2005 FDA Model Food Code provides a visual explanation to dividing your menu into no-cook, same-day, and complex categories based on the number of complete trips through the TDZ:

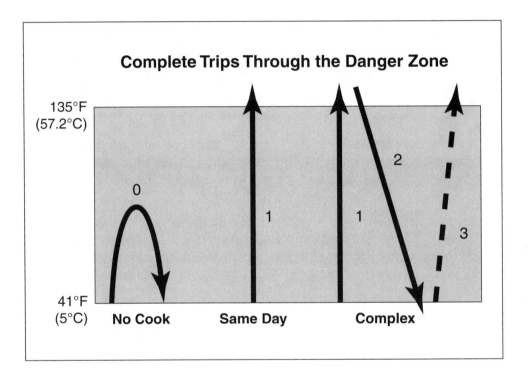

STAR KNOWLEDGE EXERCISE: MENU ITEM CATEGORIES

Are these menu items NO-COOK, SAME-DAY, or COMPLEX?

Banana Smoothie

Chicken Pasta Primavera

Chili

Mixed-Fruit Crisp

Pork Barbeque Sandwich

Quick Tuna Spread

Spanish Rice Bake

Tuna Melt

Stage 2: Hazard Evaluation

F. Analyze two questions:

★ "What is the likelihood for a hazard to occur here?"

★ "What is the risk if the hazard does occur?"

What is the likelihood for a hazard to occur?

Most hazards are biological (bacteria, viruses, and parasites). Bacterial pathogens are the most common cause of foodborne illnesses. Most of these hazards can be controlled by adequate cooking, cooling, and storage of potentially hazardous foods. Other hazards might be chemical (cleaning products, sanitizers, and pesticides) or physical (metal shavings, foreign objects, and hair).

In the tuna salad, hamburger, and chicken soup, what is the likelihood for this to occur? What are the chances these hazards could contaminate the food? Is a preventative measure needed to reduce the hazard to a safe level?

What is the risk if the hazard does occur?

If there is a **high** risk of contamination, it leads to an unacceptable heath risk that is life threatening. An **unacceptable health risk** can lead to injury, illness, or death. If the risk of contamination is **low,** it is an acceptable heath risk that is moderate or mild. Acceptable risks are those that present little or no chance of injury, illness, or death. The FDA recommends that a scientist should assist your company in analyzing the hazards and completing the risk assessments. Experts use scientific data to determine if the risk is high, medium, or low in each hazard that is analyzed. In April 2004, the FDA updated these hazards associated with food. The table that follows is a risk severity assessment based on the rating system of high, medium, or low.

High Severe Hazards	Medium Moderate Hazards: Potential Extensive Spread	Low Moderate Hazards: Limited Spread
Clostridium botulinum	Listeria monocytogenes	Campylobacter jejuni
Salmonella typhi	Salmonella spp.	Bacillius cereus
Shigella dysenteriae	Shigella spp.	Staphylococcus aureus
Hepatitis A	E. coli 0157:H7	Clostridium perfringens
Vibrio cholerae 01	Norwalk virus group	Vibrio cholerae, non-01
Vibrio vulnificus	Rotavirus	Yersinia
Trichinella spiralis	Streptococcus pyogenes	Giardia

This Star Knowledge Exercise provides an overview of the process used in the hazard analysis.

STAR KNOWLEDGE EXERCISE: HAZARD ANALYSIS

Using the foodborne illness charts provided above, identify one or two potential hazards. Rate the hazard risk as low, medium, or high. The responses may be different foodservice operations.

Sample Menu Item	Hazard(s)	Risk Assessment
Banana Smoothie		
Chicken Pasta Primavera		
Chili		
Mixed-Fruit Crisp		
Pork Barbeque Sandwich		
Quick Tuna Spread		
Spanish Rice Bake		
Tuna Melt		

The 2005 Model Food Code states that upon the completion of the hazard analysis, a list of significant hazards must be considered in the HACCP plan, along with any control measure(s) that can be used to prevent, eliminate, or reduce the hazards to a safe level. As mentioned earlier, these measures, called **control measures**, are actions or activities that can be used to prevent, eliminate, or reduce a hazard. Some control measures are not essential to food safety, while others are. Control measures essential to food safety, such as proper cooking, cooling, and refrigeration of ready-to-eat, potentially hazardous foods (time/temperature control for safety foods) are usually applied at critical control points (CCPs) in the HACCP plan (discussed later). The term *control measure* is used because not all hazards can be prevented, but virtually all can be controlled. More than one control measure may be required for a specific hazard. Likewise, more than one hazard may be addressed by a specific control measure, such as proper cooking. Here is a chart to illustrate examples of hazards and control measures for same-day service menu items.

EXAMPLES OF HOW THE STAGES OF HAZARD ANALYSIS ARE USED TO IDENTIFY AND EVALUATE HAZARDS*				
Hazard Analysis Stage		**Frozen cooked beef patties produced in a manufacturing plant**	**Product containing eggs prepared for food service**	**Commercial frozen precooked, boned chicken for further processing**
Stage 1 Hazard Identification	*Determine potential hazards associated with product*	Enteric pathogens (i.e., E. coli 0157:H7 and Salmonella)	Salmonella in finished product.	Staphylococcus aureus in finished product.
Stage 2 Hazard Evaluation	*Assess severity of health consequences if potential hazard is not properly controlled.*	Epidemiological evidence indicates that these pathogens cause severe health effects including death among children and elderly. Undercooked beef patties have been linked to disease from these pathogens.	Salmonellosis is a foodborne infection causing a moderate to severe illness that can be caused by ingestion of only a few cells of salmonella.	Certain strains of S. aureus produce an enterotoxin that can cause a moderate foodborne illness.

(continues)

EXAMPLES OF HOW THE STAGES OF HAZARD ANALYSIS ARE USED TO IDENTIFY AND EVALUATE HAZARDS* (CONTINUED)

Hazard Analysis Stage		Frozen cooked beef patties produced in a manufacturing plant	Product containing eggs prepared for food service	Commercial frozen precooked, boned chicken for further processing
Stage 2 Hazard Evaluation (continued)	*Determine likelihood of occurrence of potential hazard if not properly controlled.*	E. coli 0157:H7 is of very low probability and salmonellae is of moderate probability in raw meat.	Product is made with liquid eggs, which have been associated with past outbreaks of salmonellosis. Recent problems with Salmonella serotype Enteritidis in eggs cause increased concern. Probability of salmonella in raw eggs cannot be ruled out. If not effectively controlled, some consumers are likely to be exposed to salmonella from this food.	Product may be contaminated with S. aureus due to human handling during boning of cooked chicken. Enterotoxin capable of causing illness will only occur as S. aureus multiplies to about 1,000,000/g. Operating procedures during boning and subsequent freezing prevent growth of S. aureus; thus the potential for enterotoxin formation is very low.
	Using information above, determine if this potential hazard is to be addressed in the HACCP plan.	The HACCP team decides that enteric pathogens are hazards for this product. **Hazards must be addressed in the plan.**	HACCP team determines that if the potential hazard is not properly controlled, consumption of product is likely to result in an unacceptable health risk. **Hazard must be addressed in the plan.**	The HACCP team determines that the potential for enterotoxin formation is very low. However, it is still desirable to keep the initial number of S. aureus organisms low. Employee practices that minimize contamination, rapid carbon dioxide freezing and handling instructions have been adequate to control this potential hazard. **Potential hazard does not need to be addressed in plan.**

*For illustrative purposes only. The potential hazards identified may not be the only hazards associated with the products listed. The responses may be different for different establishments.
Source: **Food and Drug Administration 2005 Model Food Code,** http://vm.cfsan.fda.gov/~comm/nacmcfp.html#app-a.

Annex 4, Table 4: Examples of Hazards and Control Measures for Same-Day Service Items		
Process 2: Preparation for Same-Day Service		
Example Products	**Baked Meatloaf**	**Baked Chicken**
Example Biological Hazards	*Salmonella* spp.	*Salmonella* spp.
	E. coli 0-157:H7	*Campylobacter*
	Clostridium perfringens	*Clostridium perfringens*
	Bacillus cereus	*Bacillus cereus*
	Various fecal-oral route pathogens	Various fecal-oral route pathogens
Example Control Measures	Refrigeration at 41°F or below	Refrigeration at 41°F or below
	Cooking at 155°F for 15 seconds	Cooking at 165°F for 15 seconds
	Hot Holding at 135°F or above OR Time Control	Hot Holding at 135°F or above OR Time Control
	Good personal hygiene (No bare hand contact with RTE food, proper handwashing, exclusion/restriction of ill employees)	Good personal hygiene (No bare hand contact with RTE food, proper handwashing, exclusion/restriction of ill employees)
RTE = ready-to-eat food		

The key to a successful HACCP program is to conduct a thorough hazard analysis. If the hazards are **not** correctly identified, the risks to your operation increase significantly and the program will not be effective.

PRINCIPLE 2: DETERMINE CRITICAL CONTROL POINTS

PRINCIPLE 2

HACCP Principle 2 is to determine critical control points by first identifying all the standard control points in the flow of food. This helps you to identify which points are more critical than others. If a critical step is not controlled, then people can get a foodborne illness. Before determining control points and critical control points, you need to have a clear understanding of what they are.

1. **Control point (CP).** This is **any** point, step, or procedure in the flow of food where biological, physical, or chemical factors can be controlled. If loss of control occurs at this point and there is only a minor chance of contamination **and** there is not an unacceptable health risk, then the control point is not critical; it is simply a control point in the flow of food. Control point examples can be operational steps such as purchasing, receiving, storing, and preparation of the food, because if loss of control occurs, there are additional operational steps that will control the hazard.

2. **Critical control point (CCP).** This is an essential step in the product-handling process where controls can be applied and a food safety hazard can be prevented, eliminated, or reduced to acceptable levels. A **critical control point** is **one of the last** chances you have to be sure the food will be safe when you serve it. This may include cooking, cooling, hot/cold holding, maintaining specific sanitation procedures, preventing cross-contamination, or ensuring employee hygiene. It is the *critical* step that prevents or slows microbial growth. Every operation is different, so critical control points will vary from one operation to another. While not every step in the flow of food will be a CCP, there will be a CCP in at least one or more steps whenever a potentially hazardous food is in the recipe. Lack of hazard control at this point could lead to an **unacceptable health risk,** which is why it is critical. Common examples of CCPs include cooking, cooling, hot holding, and cold holding of ready-to-eat potentially hazardous foods (time/temperature control for safety of foods). Due to vegetative and spore- and toxin-forming bacteria that are associated with raw animal foods, it is apparent that the proper execution of control measures at each of these operational steps is essential to prevent or eliminate food safety hazards or reduce them to acceptable levels.

■ CRITICAL CONTROL POINT GUIDELINES

To identify critical control points, ask the following important questions. If you can answer **yes** to all these questions at any stage in the preparation process, you have identified key critical control points.

★ Can the food you are preparing become contaminated?

★ Can contaminants multiply at this point?

★ Does this step eliminate or reduce the likely occurrence of a hazard to an acceptable level?

★ Can you take corrective action(s) to prevent this hazard?

★ Is this the last chance you have to prevent, eliminate, or reduce hazards before you serve the food item to a customer?

■ DECISION TREES TO DETERMINE CRITICAL CONTROL POINTS

To identify the critical control points of a food item, we must first identify the most critical food safety procedures. The Codex Alimentarius Commission recommends using the decision tree tool to determine if the operational process is or is not a critical control point. The HACCP team can use the decision tree to give each operational step a complete evaluation to determine if it is a critical control point.

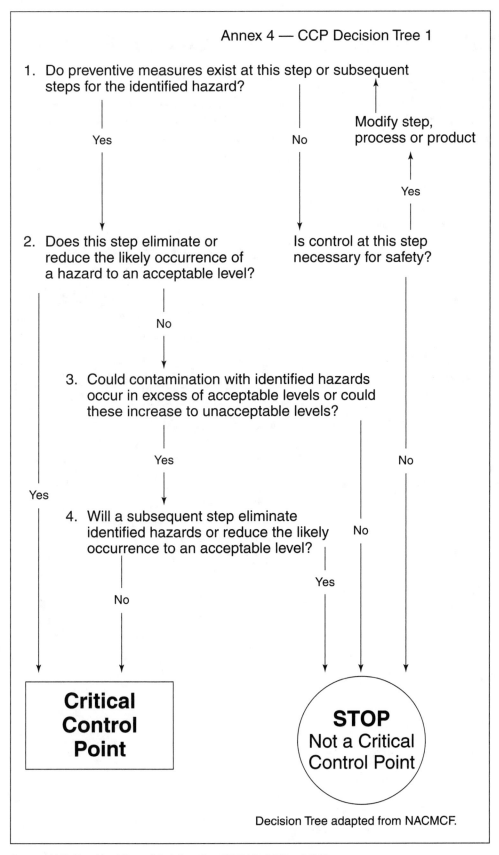

Annex 4 — CCP Decision Tree 1

1. Do preventive measures exist at this step or subsequent steps for the identified hazard?

Yes

No

Modify step, process or product

Yes

2. Does this step eliminate or reduce the likely occurrence of a hazard to an acceptable level?

Is control at this step necessary for safety?

No

3. Could contamination with identified hazards occur in excess of acceptable levels or could these increase to unacceptable levels?

Yes

No

Yes

4. Will a subsequent step eliminate identified hazards or reduce the likely occurrence to an acceptable level?

No

No

Yes

No

Critical Control Point

STOP Not a Critical Control Point

Decision Tree adapted from NACMCF.

Source: U.S. Food and Drug Administration 2005 Model Food Code.

STAR KNOWLEDGE EXERCISE: CRITICAL CONTROL POINTS

Flowchart Examples of Identifying Control Points (CP) and Critical Control Points (CCP)

Our menu items have been divided into no-cook/same-day/complex categories. In each category, review the steps in the flow of food, and then determine if each step is either a control point (CP) or a critical control point (CCP). Banana Smoothie and the Pork Barbeque Sandwich have been done for you.

No-Cook Recipes

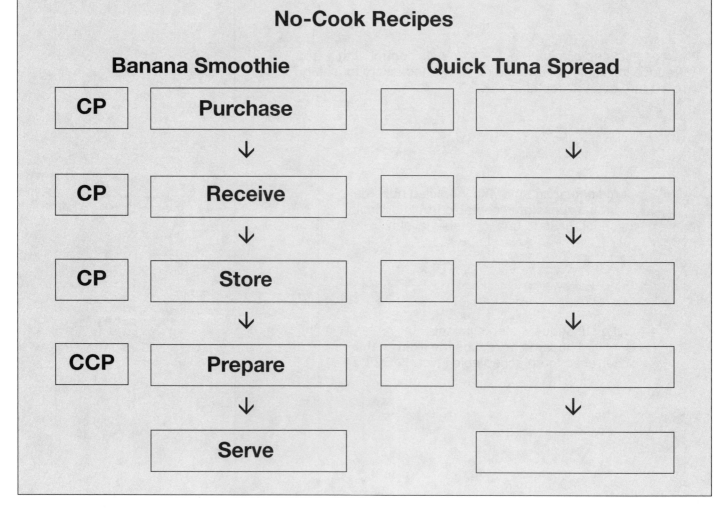

Banana Smoothie		Quick Tuna Spread	
CP	Purchase		
CP	Receive		
CP	Store		
CCP	Prepare		
	Serve		

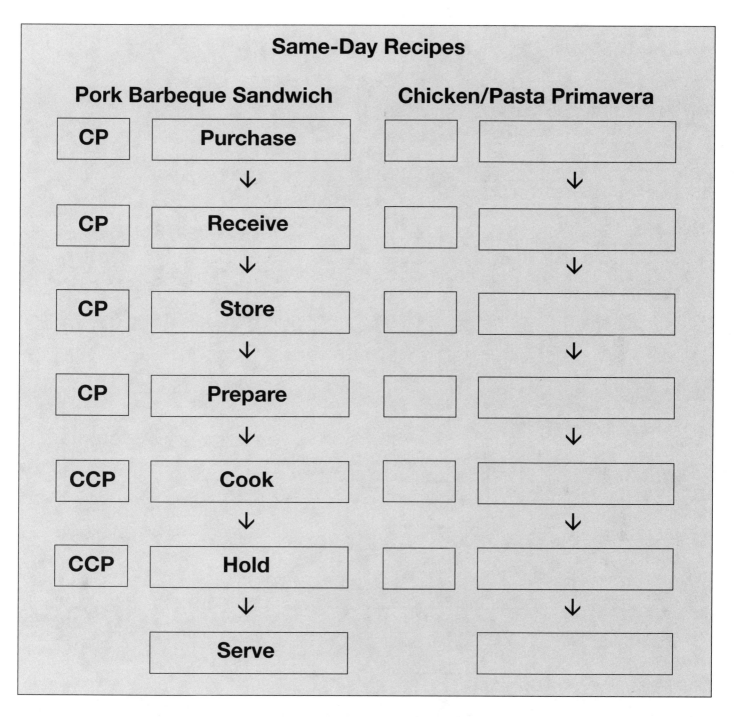

Same-Day Recipes

Pork Barbeque Sandwich

CP	Purchase
CP	Receive
CP	Store
CP	Prepare
CCP	Cook
CCP	Hold
	Serve

Chicken/Pasta Primavera

Complex Recipes

Chili

Mixed-Fruit Crisp

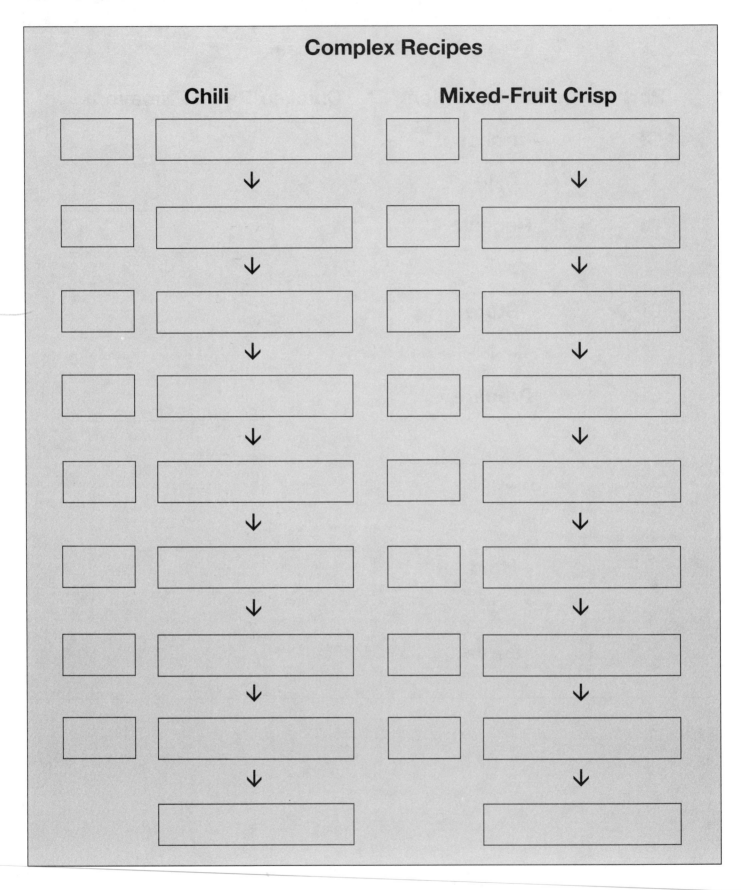

In this Star Point, we discussed how to begin setting up a HACCP plan and how to identify the hazards associated with preparing food. After learning how to identify these hazards—which is HACCP Principle 1—we learned to divide a menu into simple, same-day, and complex recipes. This then laid the foundation to understand the flow of food, control points, and critical control points—HACCP Principle 2. To better assess your knowledge of these concepts, please answer the following questions in this Star Knowledge Exercise:

STAR KNOWLEDGE EXERCISE: HACCP PRINCIPLES 1 AND 2

1. Any step in the food flow process where a hazard can be controlled is called
   a. A critical control point
   b. HACCP
   c. A control point
   d. The temperature danger zone

2. HACCP stands for
   a. Hazard Associated with Cooking Chicken Products
   b. Hazard Analysis and Critical Control Point
   c. Hazard Analysis and Control Critical Points

3. HACCP focuses on
   a. Food quality
   b. Food safety
   c. Both A and B
   d. Neither A nor B

4. A critical control point is
   a. A point where you check invoices
   b. One of the last steps where you can control, prevent, or eliminate a food safety hazard
   c. The point where food is cooled and then reheated

5. Which of the following is NOT a PHF?
   a. Foil-wrapped baked potato
   b. Chocolate chip cookie
   c. Chicken noodle soup
   d. Cut watermelon

6. What is the complete, correct sequence of the flow of food?
   a.
   b.
   c.
   d.
   e.
   f.
   g.
   h.
   i.

7. What is/are the critical control point(s) for frozen chicken nuggets served to school children?

    a. Purchase
    b. Store
    c. Cook
    d. Hold
    e. Serve

8. What type of recipe category are frozen chicken nuggets?

    a. Simple
    b. Same-day
    c. Complex

9. What type of hazard is the greatest concern to food operations?

    a. Biological
    b. Chemical
    c. Physical

10. What is the risk assessment of staphylococcus aureus on a chicken nugget?

    a. Low
    b. Medium
    c. High

Upon mastering the knowledge of the first two HACCP principles, you have successfully completed the third point of the HACCP Star.

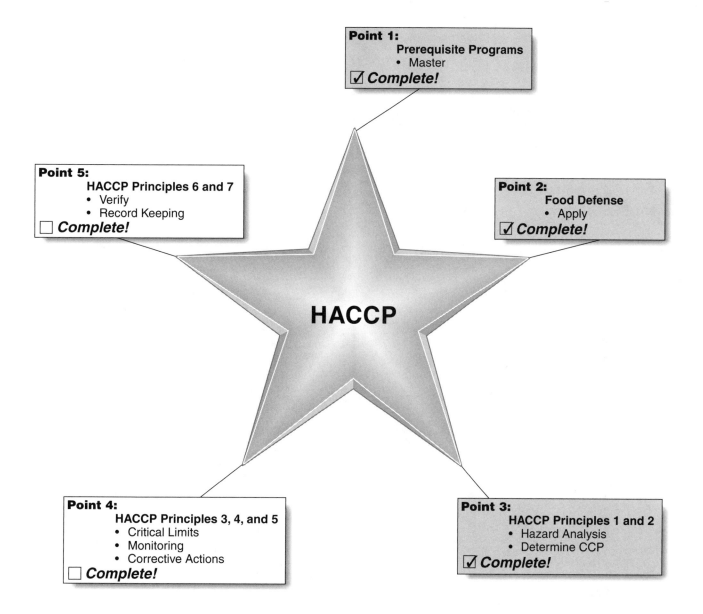

Point 1:
Prerequisite Programs
- Master

☑ *Complete!*

Point 5:
HACCP Principles 6 and 7
- Verify
- Record Keeping

☐ *Complete!*

Point 2:
Food Defense
- Apply

☑ *Complete!*

HACCP

Point 4:
HACCP Principles 3, 4, and 5
- Critical Limits
- Monitoring
- Corrective Actions

☐ *Complete!*

Point 3:
HACCP Principles 1 and 2
- Hazard Analysis
- Determine CCP

☑ *Complete!*

Work the Plan

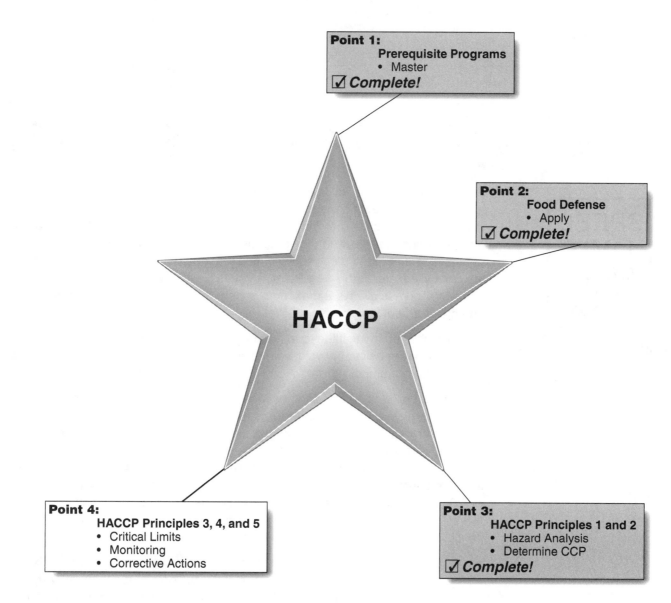

Point 1:
Prerequisite Programs
- Master

☑ *Complete!*

Point 2:
Food Defense
- Apply

☑ *Complete!*

HACCP

Point 4:
HACCP Principles 3, 4, and 5
- Critical Limits
- Monitoring
- Corrective Actions

Point 3:
HACCP Principles 1 and 2
- Hazard Analysis
- Determine CCP

☑ *Complete!*

In Star Point 4, we will discuss HACCP Principles 3 (Critical Limits), 4 (Monitoring), and 5 (Corrective Actions). This is the **most important** Star Point for your employees because this is where they "work the plan" by knowing and understanding the critical limits, then monitoring the critical limits. Finally, when a critical limit is not met, your employees must take the appropriate corrective actions to keep food safe.

Star Point Actions: You will learn to

★ Establish critical limits (HACCP Principle 3).

★ Define monitoring procedures (HACCP Principle 4).

★ Determine corrective action (HACCP Principle 5).

PRINCIPLE 3: ESTABLISH CRITICAL LIMITS

SPEED LIMIT 70

PRINCIPLE 3

Now that we have completed the hazard analysis and identified control points and critical control points, the next step is to look at critical limits. A **critical limit** is the specific scientific measurement that must clearly indicate what needs to be done and must be met for each critical control point. Critical limits must be attainable and realistic, and they must be met for each preventive measure. If the critical limit is not met, a corrective action must be performed. Corrective actions are discussed later in this Star Point. A critical limit is like a speed limit on a major highway. There is always a maximum speed. If you exceed the maximum speed and get stopped, you receive a ticket.

Critical limits determine if the food is safe or unsafe, and depending on your operation, they may consist of physical dimensions, sensory information, pH, water activity, time, and temperature.

Manufacturers, processors, and packagers of food will additionally use critical limits to determine acceptable standards for salt concentration, preservatives, free available chlorine, viscosity, humidity, pH, titratable acidity, moisture, water activity a_w, time, and temperature. Regulators take into consideration that a combination, or interaction, of two or more of these measurements may affect the stability and shelf life of processed foods. For example, semidry sausage may have a pH level of 5.0 (neutral and potentially hazardous), but its moisture-to-protein ratio (MPR) is 3.1:1, which makes it dry enough to be safe. Other factors manufacturers have critical limits for are as follows:

★ **Salt concentration.** Sodium nitrite is used in a preblended brine mix (salt, sugar, .81 percent sodium nitrite) or a combined salt/nitrite form (93.75 percent salt; 6.25 percent sodium nitrite). This is often referred to as "pickled" or "wet salted." This is a sufficient salt concentration in the food to inhibit the growth of microorganisms.

★ **Preservatives.** Sugar and salt are preservatives used in foodservice to control pathogens. Lactic acid is also used as a preservative in processing retail meats by controlling bacterial growth; it also produces a tangy flavor.

★ **Free available chlorine.** This is the concentration of hypochlorous acid (HOCL) and hypochlorite ions (OCL) existing in chlorinated water. FSIS (Food Safety and Inspection Service) regulates the use of chlorine to treat either potable water or reuse water as it enters the poultry carcass prechiller and chiller. Potable water used to initially fill the prechiller,

chiller, or red water system, or that is added as makeup water, may contain up to 50 ppm free available chlorine measured at intake and still be safe to use when preparing food.

★ **Viscosity.** A viscometer measures **viscosity**, which is the measurement in units called poise (known as centipoise) of the resistance to a change in shape. All liquids resist change in form or shape. A viscosity is the result of a fluid resistant to this change. Margarine is so viscous it seems like a soft solid. To give a clearer understanding of the centipoise, here are the approximate viscosities of a few common substances:

Substance	Approximate Viscosity at Room Temperature in Centipoise
water	1
olive oil	100
honey	2,000–10,000
molasses	5,000–10,000
Heinz ketchup	50,000–70,000
Peanut butter	250,000

Source: International Union of Pure and Applied Physics
Document U.I.P. 11 (S.U.N. 65-3).

★ **Titratable acidity.** Measuring all the various acids present. An example is the amount of alkali required to neutralize the components of a given quantity of milk and dairy products, and is expressed as percentage of lactic acid. A titration kit is used to determine the milk quality and to monitor the progress of fermentation in cheese and fermented milks.

★ **Moisture.** The USDA has **moisture-to-protein ratio** (MPR) requirements for sausage products to control the amount of added water. If too much water is present, a dangerous environment for microbial growth results and the food is no longer safe to eat. Here are some USDA MPR requirements:

  ★ Jerky—0.75:1
  ★ Pepperoni—1.6:1
  ★ Italian sausage—1.9:1
  ★ Chipped beef—2.04:1

Note: To be shelf-stable—Sausage 3.1:1 or less and a pH of 5.0 or less.

★ **Water activity (a_w).** This is the measurement of the free moisture in a food. For example, water activity is controlled in dehydrated foods, nuts, and dry mixes to extend the products' shelf life and, most importantly, to prevent the growth of dangerous microorganisms.

★ **Humidity. Relative humidity (RH)** refers to the availability of water in the atmosphere around the food or beverage. An example of using humidity as a critical limit is in the aging of beef. The dry aging of beef requires a controlled environment of humidity, time, and temperature to tenderize the beef and enhance the flavors. In addition to the aging of beef, a hygrometer is used in the measurement of humidity for ovens and dry storage areas.

For operators of food establishments, the 2005 Food Code emphasizes critical limits for time and temperature, pH, water activity, sensory information, and physical dimensions.

★ **Physical dimensions.** Hamburger patty thickness: ____in. (cm). The physical dimension and product specifications are directly related to standard operating procedures. If the supplier delivers extra-thick hamburger patties and your employees follow the current procedures for cooking them, your customers who order these burgers are at risk for E. coli 0157:H7, since the thickness of the hamburger patties is thicker than normal, unless your product is precooked prior to delivery.

★ **Sensory information.** The sensory tools used in evaluating critical limits are your nose for detecting odors, eyes for visual confirmation of quality, and hands to touch non-RTE food. For instance, if a recipe calls for fresh fish, touching the texture to feel for sliminess or stickiness will help to determine whether the fish is still fresh enough to be cooked. The smell of the fish can help to determine if it is safe to prepare. Does the fish smell fishy or does it smell like the ocean? The visual appearance of the fish is also an indication of whether or not the fish is safe to use. Are the gills red (good), or are the eyes dull, sunken in, and cloudy (bad), or are they clear and full (good)? This sensory information is necessary to determine if the fish is safe to serve to customers or not.

★ **pH.** Acidified or acid foods have a pH of 4.6 or below. For example, cooked white rice has an acidity of 6.00 to 6.70, making it neutral on a 14-point acidic scale, which is an ideal range for microorganisms (especially Bacillus cereus) to grow. By simply adding rice vinegar to the cooked rice, the acidity is lowered to 4.6 or below, deterring the growth of microorganisms. Monitoring this process with a pH meter ensures that this sushi rice is a non-potentially hazardous food. Additional examples of acidifiers are acetic acid, citric acid, lactic acid, tartaric acid, and phosphoric acid.

★ **Time and temperature.** The most common critical limit that is used in schools, contract foodservices, and independent and franchise operations is controlling time and temperature. Proper equipment and operational steps will help to prevent, reduce, or eliminate pathogens in the flow of food. A clock with a second hand or a timer and calibrated, cleaned, and sanitized thermometers are the tools needed to measure time and temperature.

■ EXAMPLES OF TIME AND TEMPERATURE CRITICAL LIMITS

Equipment

★ Conveyor belt time—Rate of heating and cooling (conveyer belt speed in): ___ft/min ___cm/min

★ Freezer temperature

★ Refrigerator temperature

★ Oven temperature: ___° F or ___°C

★ Broiler temperature: ___°F / ___°C

★ Cold-holding unit

★ Hot-holding unit

Operational Steps

★ **Receiving of products.** Meat, fish, poultry, and dairy are received at 41°F (5°C); crustacean and shellfish 45°F (7.2°C) and alive; and fresh shell eggs ambient 45°F (7.2°C).

★ **Storing of products.** Meat, fish, poultry, and dairy are stored at 41°F (5°C); crustacean and shellfish 45°F (7.2°C) and alive; and fresh shell eggs ambient 45°F (7.2°C).

★ **Preparation of items.** The maximum amount of time most food can spend in the TDZ is 4 hours. This is cumulative, combining all times in the TDZ throughout the flow of food (receiving, storing, prep, etc.) until the food is thoroughly cooked. At that point, the 4-hour time frame begins again.

★ **Cooking items.** See the Critical Limits: Minimum Cooking Temperatures Chart on pages 54 and 55.

★ **Cooling products.** Quickly cool from 135°F to 70°F (57.2°C to 21.1°C) within 2 hours, and from 70°F to 41°F (21.1°C to 5°C) within 4 hours, for a maximum total of 6 hours. Cool food as quickly as possible.

★ **Reheating products.** Reheat to 165°F (73.9°C) for 15 seconds within 2 hours.

★ **Holding items.** Hold cold food cold at 41°F (5°C) or below; hold hot food hot at 135°F (57.2°C) or above.

★ **Set up, assembly, and packing.** This may involve wrapping food items, assembling these items onto trays, and packing them into a transportation carrier or a display case. Again, the maximum amount of time most food can spend in the TDZ is 4 hours.

★ **Serving and selling.** The last operational step before the food reaches the customer is serving and selling the food. Employee's good personal hygiene and proper hand-washing procedures must be implemented with the water temperature for hand washing at 100°F (37.7°C) for 20 seconds.

If cooking is identified as the critical control point, the potentially hazardous food must meet a **minimum internal cooking temperature** that must be reached and held for 15 seconds to ensure that it is safe and does not make anyone sick.

Remember, if a recipe for baked chicken says "cook until done" or "cook until juices run clear," how do we know if the product is really safe to eat? The correct critical limit should be "cook to an internal temperature of 165°F (73.9°C) for 15 seconds." Why? Scientific data from the FDA Model Food Code provides us with documentation that proves our customers won't get sick if chicken is cooked to this designated temperature for the specific amount of time.

Here is a familiar list of cooking critical limits to help reinforce the important minimum times and temperatures needed to destroy the pathogens on food or, at minimum, to reduce them to a safe level in order to avoid foodborne illness or, worse, an outbreak. As the leader of your foodservice operation, it is essential that you emphasize the importance of the critical limits, particularly cooking times and temperatures for various food preparations to your team, to ensure that all the food served is safe for your customers to eat.

■ CRITICAL LIMITS: MINIMUM INTERNAL TEMPERATURES

165°F (73.9°C) for 15 Seconds

★ All leftover foods and reheated foods

★ All poultry and wild game

★ All stuffed products, including pasta

★ All foods cooked in a microwave; then let sit for 2 minutes

★ When combining already cooked and raw PHF products (casseroles)

155°F (68.3°C) for 15 Seconds

★ All ground foods: fish, beef, and pork

★ All flavor-injected meats

★ All eggs for hot holding and later service (buffet service)

145°F (62.8°C) for 15 Seconds

★ All fish and shellfish

★ All chops/steaks of veal, beef, pork, and lamb

★ Fresh eggs and egg products for immediate service

★ Cook roasts to 145°F for **4 minutes**

135°F (57.2°C)

★ RTE foods

★ Fully cooked commercially processed products

★ Cook or hold vegetables, and fruits

★ Hot holding for all PHF

Critical limits are the scientific and measurable standards that **must** be met. Corrective action, which means making a correction or rejection, is required if any product deviates from the critical limit. Are you able to identify critical limits in menu items? Complete the exercise in the space provided to test your knowledge.

STAR KNOWLEDGE EXERCISE: CRITICAL LIMITS

1. If holding is the CCP for tuna spread, what is the critical limit? _____

2. If cooking is the CCP for chicken pasta primavera, what is the critical limit? _____

3. If holding is the CCP for chili, what is the critical limit? _____

4. If cooking is the CCP for Spanish rice bake, what is the critical limit? _____

5. If reheating is the CCP for chili, what is the critical limit? _____

6. If cooling is the CCP for mixed-fruit crisp, what is the critical limit? _____

7. If holding is the CCP for banana smoothie, what is the critical limit? _____

8. If storage is the CCP for mixed-fruit crisp, what is the critical limit? _____

9. If cooking in the microwave is the CCP for chili, what is the critical limit? _____

10. If cooking is the CCP for a pork chop, what is the critical limit? _____

PRINCIPLE 4: ESTABLISH MONITORING PROCEDURES

Monitoring procedures, the foundation for HACCP Principle 4, ensure that we are meeting established critical limits. **Taking measurements** (of pH, water activity [a_w], time, and temperature) and **observing** are two typical tasks of monitoring. Measurements involve using an appropriate tool to get an accurate reading. Visual observation occurs when you look carefully at such products as frozen foods that must be frozen, and ensure that they show no signs of previous temperature abuse such as ice crystals and stains. There is a critical limit for every critical control point, and this must be met. If you do not check the CCPs of your food operation regularly, your HACCP plan can easily fail.

PRINCIPLE 4

Monitoring enables the manager to determine if the team is doing its part to serve and sell safe food. Monitoring also helps to **identify problems** in your foodservice operation. Monitoring provides tracking for your food safety management system throughout the operation. These concerns vary from faulty equipment (like refrigeration that will not maintain temperature), to training deficiencies (employees who continually make errors), or product specification issues (receiving goods that are not as ordered). This is by far the area in which you can shine the most as a HACCP team member—you make the difference in terms of whether or not your operation is serving safe food.

▨ HOW DO YOU MONITOR?

It is important to know your role and the roles of other team members. Namely:

★ Who will monitor the CCP(s)? Managers could, but perhaps a different team member would be better.

★ What equipment and materials are needed to monitor? Thermometers, test strips, logs, clipboards, pens.

★ How will the CCP(s) be monitored? What is the SOP? Precise directions must be given.

★ Is every team member following SOPs? Everyone, no matter what his or her job description, has duties that result in food safety.

★ When should monitoring take place? It should be built into schedules.

★ How often should monitoring take place? Some monitoring is necessary on a continuous basis or intermittently—for example, every 2 hours or every 4 hours, once per shift, or once daily or weekly.

★ Is there an accurate and permanent record of the previous monitoring that can be used for future verification? You cannot determine deviations or patterns without maintaining a log of results.

As touched on in the preceding list, there are two kinds of monitoring:

★ Continuous

★ Intermittent (noncontinuous)

Continuous monitoring is preferred because it is a constant monitoring of a CCP. This is done with built-in measuring equipment that records pH measurement, or time and temperatures. Computerized equipment systems are an example of continuous monitoring.

Intermittent (noncontinuous) monitoring occurs at scheduled intervals or on a per-batch basis and is performed often enough to ensure critical limits are met. This type of monitoring is primarily what the majority of foodservice operations use. An example of intermittent monitoring is using a properly calibrated, cleaned, and sanitized thermometer to measure the temperature of chicken soup every 2 hours.

Be a Monitoring Star

★ **Assign appropriate staff to monitor items.** The appropriate person would have the most contact with the food to be monitored. It might be a manager, but it might also be the chef who is preparing that food. It may even be the employee in charge of the salad bar. These individuals must be properly trained in using the monitoring equipment designated to measure the critical limits and in the monitoring standard operating procedures, and they should ensure the results are reported accurately. If monitoring food items and procedures is not an assigned duty, you run the risk that no one has taken it upon him- or herself to monitor the food. If no one has taken the initiative to do so, it is likely that the food is unsafe.

★ **Know how to use monitoring tools—thermometers, visual observations, and so on.** Taking the temperature of food is useless if the thermometer is not properly calibrated for accuracy, then cleaned and sanitized to prevent cross-contamination or if the thermometer has not been properly placed in the food for an appropriate amount of time. Different foods require different measuring devices. Monitoring the *sell by*, *consume by*, or *discard by* dates is an example of visual inspection or observation. Proper training is required for food employees to perform these functions effectively.

★ **Know the proper temperatures.** The reason for monitoring is not just to get a temperature reading, but to make sure the product is safe. It is a good idea to have critical limits printed on monitoring logs, recipes, posters, and job aids so the employee preparing and/or cooking the food has the standards right in front of him or her.

★ **Know ALL critical limits.** All foods have critical limits to meet at each point in the flow of food, and they don't consist of cooking limits only. As we discussed earlier in this chapter, there are other critical limits to monitor as well. Training, charts, or other means of communicating these standards to employees are important.

★ **Record monitoring results in logs.** Recording the results in logs provides the documentation necessary for verification, evaluation, adjustment, and, possibly, training and counseling employees if standards have not been met. These logs will help you to better manage your employees and your food!

★ **Perform scheduled and random monitoring tasks.** (For example, every 2 hours or every 4 hours, at a minimum) Consistent monitoring of food items definitely keeps them safe. However, there are also benefits to random checks. Checking items at precisely the same time every day may not alert you to fluctuations in equipment that could affect food safety. Varying monitoring schedules may allow food employees to gain more monitoring experience and expertise regarding critical limits, which will make them even more valuable to you.

Use Monitoring Forms

In the HACCP system, proper documentation must be maintained throughout the operational food flow. Effectively using monitoring forms when receiving, preparing, cooking, cooling, reheating, and storing a food item provides a product history. This verifies that a product meets standards or indicates when adjustments to the system are needed. It also provides documentation that you have done all you can do to keep the food safe if a foodborne illness outbreak occurs. These records provide your documentation to the Department of Health; the media; and local, county, state, and federal food inspectors.

Equipment temperatures during meal preparation and service should be monitored at least every 4 hours; this includes all refrigeration, cooking, and holding equipment. If necessary, adjust the equipment thermostats so products meet the required temperature standards.

Every foodservice operation should establish standard operating procedures for specific documentation when monitoring. These SOPs include sanitation practices, employee practices, and employee training. Each of these SOPs may consist of an informal notation of observations concerning what is working well and what is not working well. This documentation helps to identify practices and procedures that may have to be modified and may indicate a need for additional employee training.

SOP: Monitoring Instructions (Sample)

1. Inspect the delivery truck when it arrives to ensure that it is clean, free of putrid odors, and organized to prevent cross-contamination. Be sure refrigerated foods are delivered on a refrigerated truck.

2. Check the interior temperature of refrigerated trucks.

3. Confirm vendor name, day and time of delivery, as well as driver's identification before accepting delivery. If the driver's name is different than what is indicated on the delivery schedule, contact the vendor immediately.

4. Check frozen foods to ensure that they are all frozen solid and show no signs of thawing and refreezing, such as the presence of large ice crystals or liquids on the bottom of cartons.

5. Check the temperature of refrigerated foods.

   a. For fresh meat, fish, and poultry products, insert a clean and sanitized thermometer into the center of the product to ensure a temperature of 41°F (5°C) or below.

   b. For packaged products, insert a food thermometer between two packages, being careful not to puncture the wrapper. If the temperature exceeds 41°F (5°C), it may be necessary to take the internal temperature before accepting the product.

   c. For eggs, the interior temperature of the truck should be 45°F (7.2°C) or below.

6. Check dates of milk, eggs, and other perishable goods to ensure safety and quality.

7. Check the integrity of food packaging.

8. Check the cleanliness of crates and other shipping containers before accepting products. Reject foods that are shipped in dirty crates.

Examples of Monitoring Forms

THERMOMETER CALIBRATION LOG: (AS SCHEDULED)

	DATE 10/21	EMPLOYEE	MGR ✓	DATE 10/22	EMPLOYEE	MGR ✓
6 AM	32 33*32 32	JS	EL			
2 PM	36*32 33 32	EL	EL			
10 PM	32 32 32	KK	MR			

COOKING LOG:

DATE	TIME	FOOD PRODUCT	INTERNAL TEMPERATURE °F or °C	CORRECTIVE ACTION TAKEN	EMPLOYEE INITIALS	MANAGER INITIALS
10/21	2 pm	Chili	120	Reheat	EL	MP

COOLING LOG: DATE: 10/21   EMPLOYEE: JS   MANAGER: FA

DATE	FOOD	TIME	TEMP	+2 HOURS TIME	MUST BE 70°F OR LOWER	+3 HOURS TIME	TEMP	+4 HOURS TIME	TEMP	+5 HOURS TIME	TEMP	+6 HOURS TIME	MUST BE 41°F OR LOWER
10/21	Chili	9 PM	112	11 PM	68	12 PM	52	1 PM	39				

HOLDING LOG: DATE: 10/21   EMPLOYEE: BG   MANAGER: ML

FOOD	6 AM	10 AM	2 PM	6 PM	10 PM	2 AM	6 AM	10 AM	2 PM	6 PM	10 PM	2 AM
Spanish rice	—	165	145	147								

STAR KNOWLEDGE EXERCISE: MONITORING

What type(s) of monitoring logs and food safety equipment would you need to have in place for the following foods? Place the appropriate letter(s) of the monitoring logs and equipment that should be used to monitor each food item described below. Are there any additional logs or food safety equipment that you would use in your foodservice operation?

Monitoring Logs

A. Receiving log or invoice (indicating temperature of frozen and/or refrigerated food as received)
B. Freezer temperature log
C. Refrigerator temperature log
D. Dry storage temperature log
E. Preparation temperature log
F. Cooling log
G. Reheating log
H. Serving line temperature log

Equipment

I. Calibrated thermometers
J. pH meter
K. Display tank
L. Refrigeration unit
M. Cold-holding equipment
N. Cold-serving equipment
O. Freezer unit
P. Oven
Q. Hot-holding equipment
R. Hot-serving equipment
S. Equipment to quickly cool soup—ice bath, ice paddle, pans to reduce product for quicker cooling; or a blast chiller

Food to Be Monitored

1. Tuna spread prepared on-site to be served that day.

2. BBQ pork to be cooked on-site and served that day.

3. Chili prepared on-site to be served the next day.

4. Potato salad prepared on-site for deli case and salad bar.

5. Prime rib cooked on-site and served that day.

6. Chicken vegetable soup prepared on-site for salad bar also served the next day.

7. Veal parmesan for catered luncheon prepared on-site to be served that day.

8. Fresh peach ice cream prepared on-site.

9. Live lobsters in display tank to be served the next day.

10. Sushi prepared on-site.

PRINCIPLE 5: IDENTIFY CORRECTIVE ACTIONS

Now that the minimum and maximum critical limits have been identified and recorded through your monitoring efforts, we can identify the corrective actions necessary to fix any deficiencies. The **corrective actions** are predetermined steps that you automatically take if the critical limits are not being met. This is Principle 5 of the HACCP Process. There are five tasks necessary when a corrective action occurs:

1. Establish the exact cause of the deficiency.
2. Determine who is to correct the problem.
3. Correct the problem.
4. Decide what to do with product.
5. Record the corrective actions that were taken.

Following are examples of corrective actions:

★ **Reject a product that does not meet purchasing or receiving specifications.** If, upon inspection when receiving a shipment, you find that a substitution has been made to the size, quantity, brand, or other issues relating to the food item, do not accept the product, since any of these inconsistencies could affect the quality of your recipe, or even compromise the safety of your product.

★ **Reject a product that does not come from a reputable source.** Risks when receiving food items from unreliable sources include poor-quality food, unsafe food, no recourse if refunds are needed, the inability to do trace backs in the event of foodborne illness outbreaks, liability, and so on.

★ **Fix, calibrate, or replace thermometers in or on refrigerators, freezers, ovens, cold-holding carts, hot-holding carts, and so on.** Otherwise, how can you ensure the food that you have paid for will remain safe?

★ **Discard unsafe food products**. If food is identified as **adulterated** (contains any poisonous or deleterious substance), it cannot be eaten or sold, and it should be disposed of so it is not mistakenly used.

★ **Discard food if cross-contamination occurs,** especially if there is no cooking step involved, since there's no possibility of destroying any pathogens that are on the food. Food that is touched or dropped or otherwise cross-contaminated will introduce any number of unknown risks to the customer. If you cannot make it safe, you cannot serve it.

★ **Continue cooking food until it reaches the correct temperature**. Monitoring this process effectively can eliminate the occurrence of undercooked food. No one can calculate the temperature of food by looking at it, so the calibrated thermometers are the tools that will enable you to ensure foods are cooked to their safe temperatures.

★ **Reheat food to 165°F (73.9°C) within 2 hours.** Monitoring will alert you to any foods that have lingered in the temperature danger zone (TDZ), thereby allowing for rapid growth of microorganisms. Quickly raising the temperature to 165°F (73.9°C) will destroy enough of the microorganisms to make the food safe for consumption. This can apply to foods that are holding on a buffet line or left out to cool.

★ **Change methods of food handling.** Monitoring by observation may indicate food employees are not following (or understanding) proper food-handling methods. Retraining might be necessary for food employees who do not wash hands between tasks or when they have touched a dirty surface. Procedures must be explained and enforced to prevent cross-contamination by washing, cleaning, and sanitizing food-contact surfaces and utensils. Proper food handling also includes practices of safely thawing foods, separating raw and RTE foods, and cleaning and sanitizing the containers, utensils, and hands that have contacted the raw and RTE foods.

★ **Train staff to calibrate thermometers and take temperatures properly.** An uncalibrated thermometer will give a false reading that may lead to serving unsafe foods. Thermometers should be calibrated every shift. Using the wrong thermometer, or one not inserted into the food properly, may also be misleading.

★ **Document: Write everything down!** Don't forget to record what you have done to correct the problems you have observed when monitoring. This practice is important and helpful if a customer is stricken with a foodborne illness. These documents provide evidence that your HACCP system has been implemented.

STAR KNOWLEDGE EXERCISE: CORRECTIVE ACTION

Using the food products from our sample restaurant menu (page 164), take a look at the critical control point, the critical limit, and the monitoring results identified. Based on the monitoring results, what would you determine the corrective action to be? Assume the item has been checked within two hours.

Product	CCP	Critical Limit	Monitoring	Corrective Action
Tuna Spread	Hold	41°F	48°F	
Chili	Hold	135°F	121°F	
BBQ	Cook	135°F	135°F	
Tuna Melt	Prep	41°F		
Spanish Rice Bake	Cook	165°F	160°F	
Chicken Pasta Primavera	Cook	165°F	145°F	
Banana Smoothie	Store milk	Exp. Date	Old	
Mixed-Fruit Crisp	Reheat	165°F	155°F	

For all SOPs, it is mandatory that predetermined steps are developed **before** the deficiency occurs so employees and managers know what to do when it occurs. Here is an example of corrective actions for receiving deliveries.

SOP: Receiving Deliveries (Sample)

Corrective Action:

1. Reject the following:

   a. Frozen foods with signs of previous thawing

   b. Cans that have signs of deterioration—swollen sides or ends, flawed seals or seams, dents, or rust

   c. Punctured packages

   d. Expired foods

   e. Foods that are out of safe temperature zone or deemed unacceptable by the established rejection policy

STAR KNOWLEDGE EXERCISE: CORRECTIVE ACTION

Using recipe items from your own establishment, complete the sample Corrective Action charts.

RECEIVING LOG: DATE:

TIME	TEMP	FOOD PRODUCT DESCRIPTION	PRODUCT CODE	CORRECTIVE ACTION TAKEN	EMPLOYEE INITIALS	MANAGER INITIALS

COOLING—CORRECTIVE ACTION LOG:

DATE	FOOD PRODUCT	TIME	TEMPERATURE MUST: 70°F (21.1°C)–2 HOURS MUST: 41°F (5°C)–6 HOURS	CORRECTIVE ACTION TAKEN • MUST: REHEAT • MUST: DISCARD	EMPLOYEE INITIALS	MANAGER INITIALS

REFRIGERATION LOG:

DATE	TIME	TYPE OF UNIT	LOCATION	°F/°C	CORRECTIVE ACTION TAKEN	EMPLOYEE INITIALS	MANAGER INITIALS

STAR KNOWLEDGE EXERCISE: HACCP PRINCIPLE CHECK

At this point, your HACCP system should be operating effectively, since you have executed the prerequisite programs and applied food defense and HACCP Principles 1 through 5. As a quick review, what are the five HACCP principles we've covered so far?

★ Principle 1: _____

★ Principle 2: _____

★ Principle 3: _____

★ Principle 4: _____

★ Principle 5: _____

Once developed, your HACCP plan is your system for self-inspection because you are responsible for executing the plan. Even though you can do self-inspections, your local regulatory official must also inspect your business. He may determine everything is fine, or he may be concerned because after a thorough inspection of your foodservice operation, he discovered some out-of-control procedures. In an effort to help get your procedures back on track and under control, he might request that you complete a **risk control plan (RCP)**. He explains that the risk control plan is a written plan created by you (the foodservice operator) with input from him (your local health inspector) that describes how to manage specific out-of-control risk factors. Then he hands you this chart from the 2005 Model Food Code with specific solutions to assist you in serving and selling safe food:

Out-of-Control Procedure	Associated Hazards	On-Site Correction (COS)	Long-Term Compliance
Bare-hand contact with RTE Food	Bacteria, parasites, and viruses via fecal-oral route	Conduct hazard analysis.	Risk control plan (RCP), train employees, SOP/HACCP development.
Cold holding	Vegetative bacteria, toxin-forming and spore-forming bacteria, scrombrotoxin (finfish)	Conduct hazard analysis.	Change equipment, RCP, train employees, develop SOP/HACCP/recipe.
Contaminated equipment	Bacteria, parasites, and viruses	Clean and sanitize equipment; discard or reheat RTE food.	Train employees, change equipment or layout, develop SOP.
Cooking	Vegetative bacteria, parasites, and possibly viruses	Continue cooking to proper temperature.	Change equipment, RCP, train employees, develop SOP/HACCP/recipe.
Cooling	Toxin-forming and spore-forming bacteria	Conduct hazard analysis.	Change equipment, RCP, train employees, develop SOP/HACCP/recipe.
Cross-contamination of RTE foods with raw animal foods	Bacteria, parasites, and possibly viruses	Discard or reheat RTE food.	Change equipment layout, RCP, train employees, develop SOP/HACCP/recipe.
Food source/sound condition	Bacteria/parasites/viruses/scombrotoxin/ciguatera toxin	Reject or discard.	Change buyer specifications, train employees.
Freezing to control parasites	Parasites	Freeze immediately; discard; or cook.	Change buyer specifications, RCP, develop SOP/HACCP/recipe, change equipment, train employees.
Hand washing	Bacteria, viruses, and parasites	Wash hands immediately; conduct hazard analysis.	Change equipment layout, train employees, RCP, develop SOP/HACCP.
Hot holding	Toxin-forming and spore-forming bacteria	Conduct hazard analysis.	Change equipment, RCP, train employees, develop SOP/HACCP/recipe.
Receiving temperatures	Scombrotoxin, bacteria	Reject or discard.	Change buyer specifications, train employees, develop SOP/HACCP/recipe.
Reheating	Vegetative bacteria; toxin-forming and spore-forming bacteria	Conduct hazard analysis.	Change equipment, RCP, train employees, develop SOP/HACCP/recipe.

Specifically, your regulator may have observed the temperature of turkey vegetable soup in your walk-in cooler was 65°F (18.3°C) after cooling in the walk-in all night for 12 hours. Now, the moment of truth has arrived, and he tells you to complete a risk control plan (RCP) for the turkey vegetable soup:

Risk Control Plan

Establishment Name: _____ Type of Facility: _____

Physical Address: _____ Person in Charge: _____

City: _____ State: _____ Zip: _____ County: _____

Inspection Time In: _____ Inspection Time Out: _____ Date: _____ Inspector's Name: _____

Agency: _____

Specific observation noted during inspection:

_____
_____
_____
_____
_____
_____
_____
_____
_____
_____

Applicable code violation(s):

_____
_____
_____
_____
_____
_____
_____
_____

Risk factor to be controlled:

_____
_____
_____
_____
_____
_____

What must be achieved to gain compliance in the future?

_____

_____

_____

_____

_____

_____

_____

_____

_____

How will active managerial control be achieved?

(Who is responsible for the control, what monitoring and record keeping is required, who is responsible for monitoring and completing records, what corrective actions should be taken when deviations are noted, how long is the plan to continue)

How will the results of implementing the RCP be communicated back to the inspector?

As the person in charge of the _____ located at _____, I have voluntarily developed this risk control plan, in consulation with _____, and understand the provisions of this plan.

_____   _____

(Establishment manager)   (Date)

_____   _____

(Regulatory official)   (Date)

Example: Risk Control Plan for Turkey Vegetable Soup

Establishment Name: ABC Establishment		Type of Facility: Full Service
Physical Address: 123 Any Street		Person in Charge: John Doe

City: Any City	State: Any State	Zip: 00000	County: Any County

Inspection Time In: 9:00 a.m.	Inspection Time Out: 12:30 p.m.	Date: 10.10.05	Inspector's Name: Jane Doe

Agency: Your jurisdiction

Specific observation noted during inspection:

Temperature of turkey vegetable soup in walk-in cooler was 65°F (18.3°C) after cooling in the walk-in all night (12 hours).

Applicable code violation(s):

Food Code Section 3-501.14—Soup not cooled from 135°F–41°F (57.2°C to 5°C) in 6 hours or less.

Risk factor to be controlled:

Improper Holding Temperatures (Cooling)

What must be achieved to gain compliance in the future?

Cool from 135°F to 41°F (57.2°C to 5°C) within 6 hours provided that food is cooled from 135°F to 70°F (57.2°C to 21.1°C) in ≤2 hours.

How will active managerial control be achieved?

(Who is responsible for the control, what monitoring and record keeping is required, who is responsible for monitoring and completing records, what corrective actions should be taken when deviations are noted, how long is the plan to continue)

(Attached is Cooling SOP)

Conduct a Trial Run to Determine If Cooling Procedure Works

The head chef will portion soup at a temperature of 135°F (57.2°C) in cleaned and sanitized 3-inch (7.62-cm) metal pans, and when the soup has reached 70°F (21.2°C), place them uncovered in the coolest, protected area of the walk-in cooler. He will record the time on the "Time-Temperature Log." Two hours later, the temperature of the soup will be checked and recorded. If the temperature of the soup is not 70°F (21.1°C) or less, the soup will be reheated to 165°F (73.9°C), and the trial run will be restarted in an ice bath. When the temperature is 70°F (21.2°C) or less within 2 hours, the time and temperature will be recorded, and cooling will continue. Four hours later, the temperature of the soup will again be checked and recorded. If the soup is 41°F (5°C) or less, the cooling procedure will be established. If the soup is not 41°F (5°C) or less, it will be discarded and other cooling options will be used (see below).

Procedure

When there is less than one gallon of soup left over at the end of the day, the head chef will log the volume and disposition of the soup. When the volume is greater than one gallon, the established procedure will be followed. The head chef will complete the Temperature Log daily for 30 days. The general manager will review the log weekly for completeness and adherence to the procedure.

> Other options that may be suggested to the operator include purchasing a data logger to record cooling overnight; discarding any leftover soup at the end of the day; using chill sticks/ice paddles; using an ice bath to cool leftovers prior to storage; and purchasing a blast chiller.

How will the results of implementing the RCP be communicated back to the inspector:

The log will be available for review by the regulatory authority upon request.

As the person in charge of the _____ located at _____, I have voluntarily developed this risk control plan, in consultation with _____, and understand the provisions of this plan.

_____   _____

(Establishment manager)   (Date)

_____   _____

(Regulatory official)   (Date)

In Star Point 4, Work the Plan, we discussed why this is the most important Star Point for you. This is where **you make the greatest difference!** You do this by "working the plan"—actively monitoring critical control points and critical limits and identifying and facilitating corrective actions that ensure safe food. The principles presented in Star Point 4 are at the crux of any HACCP plan and determine its success when the plan is put into action. If your facility is asked to complete an RCP, it is critical that you perform verification of your prerequisite programs and your entire HACCP plan. This verification leads to the next Star Point.

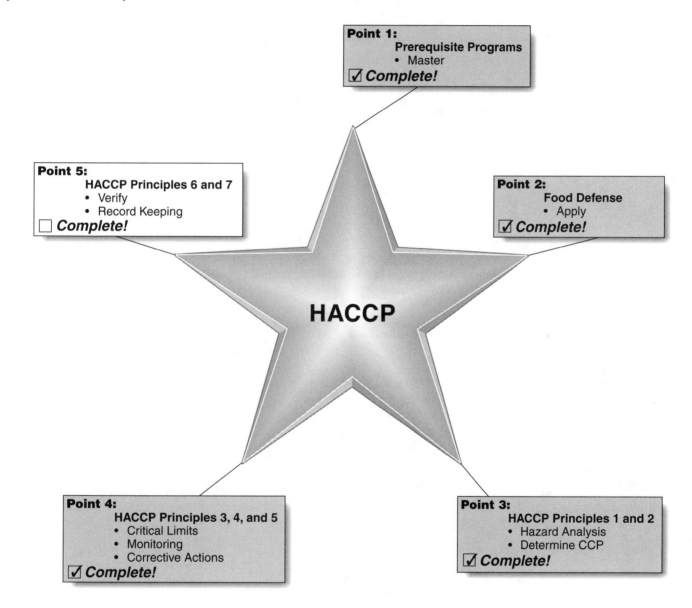

Checks and Balances

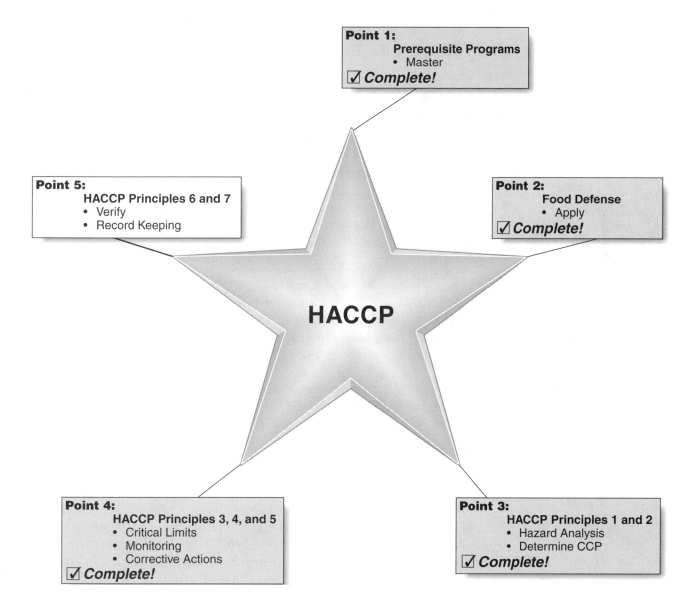

Point 1:
 Prerequisite Programs
 • Master
 ☑ *Complete!*

Point 5:
 HACCP Principles 6 and 7
 • Verify
 • Record Keeping

Point 2:
 Food Defense
 • Apply
 ☑ *Complete!*

HACCP

Point 4:
 HACCP Principles 3, 4, and 5
 • Critical Limits
 • Monitoring
 • Corrective Actions
 ☑ *Complete!*

Point 3:
 HACCP Principles 1 and 2
 • Hazard Analysis
 • Determine CCP
 ☑ *Complete!*

In this final Star Point, we discuss the checks and balances system of your HACCP plan, which consists of HACCP Principles 6 (Verification) and 7 (Record Keeping). **Verification** enables your plan to be confirmed if everything is working properly or modified as necessary to conform to the appropriate standards. **Record keeping**, or documentation, is written proof that your HACCP system is in action and that you prepare, serve, and sell safe food. If a foodborne illness or outbreak were to occur, HACCP records provide a reasonable care defense for your foodservice operation.

Star Point Actions: You will learn to

★ Confirm that the HACCP system works (HACCP Principle 6)

★ Maintain records and documentation (HACCP Principle 7)

PRINCIPLE 6: VERIFY THAT THE HACCP SYSTEM WORKS

PRINCIPLE 6

There are only two principles that remain in the HACCP plan. Verification, which is Principle 6, is a check or a **confirmation** that the steps of the plan are working. Verification ensures that your operation is maintaining an effective food safety management system, and it provides a chance to update the plan as needed. Verification activities are always completed by the supervisor, director, or even an outside firm.

A component of verification is validation, which must be executed by a food safety professional if HACCP is required. Federal or state regulatory officials, qualified food safety consultants, or members of academia should complete this process. If a foodservice facility executes a food safety management system committed to active managerial control, validation is voluntary and can be performed in-house. According to the 2005 Food Code, **validation** focuses on collecting and evaluating scientific and technical information to determine if the HACCP system, when properly implemented, will effectively control the hazards. (On pages 218–219 is an example of a validation worksheet to be used by the HACCP team.)

When verification takes place the following questions must be answered:

★ **Who** will perform the verification? It is critical to remember that the person responsible for performing the verification process should reinforce management's commitment to HACCP as well as the prerequisite programs and achieving active managerial control of foodborne illness risk factors. The leadership impact of verification is very powerful, so make sure your company chooses someone who is responsible, credible, and has a strong commitment to food safety, your organization, and your brand. This person confirms without a doubt that the HACCP plan in place is working to prevent death or injury to you, your employees, your customers, or your facility.

★ **What** needs to be performed? Verification requires continuous review of logs, records, corrective actions, and so on to understand whether or not your HACCP plan is working. The verification process requires that a procedure be developed with specific objectives and steps to effectively determine if the HACCP plan is working the way it was designed to work. Verification assessment is not a guessing game; it is the reality of checks and balances when **reviewing** the required logs, charts, and records. It validates that employees were trained to maintain SOPs and prerequisite programs. Verification is based on the actual performance of management once a HACCP plan is in place. For this reason, it is necessary to implement a HACCP plan so that the required behavior is part of the employees' routine duties, rather than treating it as something out of the ordinary. The HACCP team and leadership in your organization have developed audits, checklists, and systems to verify that the HACCP plan, including prerequisite programs and CCPs, are correct.

★ **Where** will this be performed? Verification procedures can be performed on-site or off-site. On-site procedures can include self-inspections. Using observations and checklists, the verification team can review the HACCP plan, observe whether or not prerequisite programs are being followed, review deviations and recommend corrective actions, take random samples for analysis, review CCP records and critical limits to verify that they are adequate to control hazards, and make any modifications to the HACCP plan, as necessary. Some off-site **validation activities** would be microbiological testing and analysis of data from the records, logs, and charts by scientists and highly skilled professionals.

★ **When** will verification be performed? Verification must be routine but unannounced to ensure a real picture of the operation and **determining** the correct frequency is based on the monitoring procedures. Prerequisite programs, including all SOPs, need to be reviewed on an annual basis. Verification must be performed on an ongoing basis, such as when new equipment is added to the kitchen, when new items have been added to the menu, and after any personnel changes have taken place. In addition, HACCP verification is required when a foodborne illness has been reported or a foodborne illness is associated with a food that your organization produces or serves. Finally, verification is needed when changes are made to the FDA Model Food Code and other regulations are modified, to ensure that you are in compliance with current codes.

★ **Why?** Verification confirms that the HACCP plan is working on a continuous basis. This process ensures that you are serving safe food and protecting your customers!

★ **How** will verification be performed? Verification is performed by using a Verification Inspection Checklist, HACCP Plan Verification Worksheet, HACCP Plan Verification Summary, or some other systematic process that **ensures** corrective actions were taken by reviewing records, logs, and charts, calibrating equipment and tools, and by observing or interviewing personnel as they do their work. These recommended tools (Verification Inspection Checklist, HACCP Plan Verification Worksheet, and HACCP Plan Verification Summary) can be modified and are located at the end of Principle 6. The verification team can also take a random sampling and compare their results with previously written records to ensure accuracy. Microbiological testing by laboratories can **validate** that the critical limits are preventing, reducing, or eliminating the hazard. Third-party audits can assist in the verification process by analyzing food samples; by checking the signature and date on records, logs, and charts; and by actually **confirming** all equipment is properly calibrated, maintained, and operating.

Validation Worksheet

Name of person responsible for validation: _____

Title: _____

Frequency at which the validation is done: _____

Reason, other than frequency, for doing the validation: _____

Date of last validation: _____

The length of time this record is kept on file (i.e. # months or years): _____

1. (a) Has a new product, process, or menu item been added since the last validation?

   No _____
   Yes _____
   Go to Question #1b

   (b) Has the supplier, customer, equipment, or facility changed since the last validation?

   No _____
   Yes _____
   Go to question #2

2. Are the existing worksheets accurate and current?

   No _____
   Yes _____ ↑
   Go to Question #3

   Worksheet information updated: Date: _____   Name: _____

3. Are the identified hazards accurate and current?

   No _____
   Yes _____ ↑
   Go to Question #4

   Hazard analysis updated: Date: _____   Name: _____

4. Are the existing CCPs correctly identified?

   No _____
   Yes _____ ↑
   Go to Question #5

   CCPs updated: Date: _____   Name: _____

5. **Are the existing critical limits appropriate to control each hazard?**

   No ——→ CLs updated: Date: _____   Name: _____

   Yes ——

   Go to Question #6

6. **Do the existing monitoring procedures ensure that the critical limits are met?**

   No ——→ Monitoring procedures updated: Date: _____   Name: _____

   Yes ——

   Go to Question #7

7. **Do existing corrective actions ensure that no injurious food is served or purchased?**

   No ——→ Corrective Actions updated: Date: _____   Name: _____

   Yes ——

   Go to Question #8

8. **Do the existing on-going verification procedures ensure that the food safety system is adequate to control hazards and is consistently followed?**

   No ——→ On-going verification procedures updated: Date: _____   Name: _____

   Yes ——

   Go to Question #9

9. **Does the existing record keeping system provide adequate documentation that the critical limits are met and corrective actions are taken when needed?**

   No ——→ Record keeping procedures updated: Date: _____   Name: _____

   Yes ——

   Go to Question #10

10. **Are the existing prerequisite programs current?**

   No ——→ Prerequisite Programs updated: Date: _____   Name: _____

   Yes ——

The validation procedure is now complete. The next validation is due _____.

The changes made to the food safety management system were conveyed to the line supervisor or front-line employees on _____.

Completed by: _____   Name _____

Source: From Managing Food Safety: A Manual for the Voluntary Use of HACCP Principles for Operators of Food Service and Retail Establishments, July 2005, Center for Food Safety and Applied Nutrition, Food and Drug Administration, U.S. Department of Health and Human Services, http://www.cfsan.fda.gov/~acrobat/hret2.pdf

The following Verification Inspection Checklist, HACCP Plan Verification Worksheet, and HACCP Plan Verification Summary are commonly used by regulatory authorities in checking all steps in the HACCP plan for accuracy and effectiveness. These are outstanding tools for foodservice HACCP teams to use when evaluating their own HACCP plan because they are logical and systematic.

FORM — Verification Inspection Checklist

Date: _____   Time: _____   Scheduled (S)/Unscheduled (U): _____

Establishment Name: _____

Establishment Address: _____

Person in Charge: _____

Document Review

1. **Documents provided for review:**

Type of Document	Reviewed (Y or N)	Comments/Stengths/Weaknesses Noted
Prerequisite Programs (list them below)		
Menu or Food List or Food Preparation Process		
Flow Diagrams (Food Preparation)		
Equipment Layout		
Training Protocols		
Hazard Analysis		
Written Plan for Food Safety Management System		
Other		

FORM (continued)

2. List critical control points (CCPs) and critical limits identified by the establishment's HACCP plan.

Food Item or Process	Critical Control Point	Critical Limits	Comments/Problems Noted

3. What monitoring records are required by the plan?

Type of Record (Prerequisite Program Activities, Monitoring, Corrective Action, CCP Verification, etc.)	Monitoring Frequency and Procedure (How often?, Initialed and dated?, etc.)	Record Location (Where kept?)

4. Describe the strengths or weaknesses with the current monitoring or record-keeping regimen.

Comments:_____

_____

5. Who is responsible for verification that the required records are being completed and being properly maintained?

Comments:_____

_____

FORM (continued)

6. **Describe the training that has been provided to support the system.**

Comments:_____

_____

_____

7. **Describe examples of any documentation that the above training was accomplished.**

Comments:_____

_____

_____

Record Review and On-Site Inspection

(Choose at random one week from the previous four.)

8. **Are monitoring actions performed according to the plan?**

☐ Full compliance   ☐ Partial compliance   ☐ Noncompliance

Comments:_____

_____

_____

9. **When critical limits established by the plan are not met, are immediate corrective actions taken and recorded?**

☐ Yes   ☐ No

Comments:_____

_____

_____

10. **Do the corrective actions taken reflect the same actions described in the establishment's plan?**

☐ Yes   ☐ No

Comments:_____

_____

_____

11. **Are routine calibrations required and performed according to the plan?**

☐ Yes   ☐ No

Comments:_____

_____

_____

FORM (continued)

---

12. **Examine the current day's records, if possible. Are the records for the present day accurate for the observed situation in the facility?**

    ☐ Yes   ☐ No

    Comments:_____

    _____

    _____

---

13. **Do managers and employees demonstrate knowledge of the system?**

    Managers: ☐ Yes ☐ No   **Employees:** ☐ Yes ☐ No

    Comments:_____

    _____

    _____

---

14. **Have there been any changes to the menu or recipes since the last verification visit?**

    ☐ Yes   ☐ No

    Comments:_____

    _____

    _____

---

15. **Was the system modified because of these menu or recipe changes?**

    ☐ Yes   ☐ No

    Comments:_____

    _____

    _____

Additional Comments or Recommendations:

FORM

HACCP Plan Verification Worksheet

(Note: This document is for optional use only, and is not a requirement for the Standardization Procedure.)

Establishment Name:	**Type of Facility:**

Physical Address:	**Person in Charge:**

City:	**State:**	**Zip:**	**County:**

Inspection Time In:	**Inspection Time Out:**	**Date:**	**Candidate's Name:**

Agency:	**Standard's Name:**	**Indicate Person Filling Out Form:** (circle one) **Candidate's Form/Standard's Form**

Cold Holding Requirement for Jurisdiction: 41°F (5°C) _____ or 45°F (7°C) _____ or 41°F (5°C) and 45°F (7°C) combination: _____

1. Have there been any changes to the food establishment menu?

Yes _____   No _____

Describe:_____

_____

_____

2. Was there a need to change the food establishment HACCP plan because of these menu changes?

Yes _____   No _____

Describe:_____

_____

_____

3. List critical control points (CCPs) and critical limits (CLs) identified by the establishment HACCP plan.

CCPs	CLs
_____	_____
_____	_____
_____	_____
_____	_____

4. What monitoring records for CCPs are required by the plan?

Type of record	Monitoring frequency	Record location
_____	_____	_____
_____	_____	_____
_____	_____	_____
_____	_____	_____

FORM (continued)

5. Record compliance under 4G of the *FDA Standardization Inspection Report* (ANNEX 2 Section 1). Are monitoring actions performed according to the plan?

   Yes _____   No _____

   Describe:_____
   _____
   _____

6. Is immediate corrective action taken and recorded when CLs established by the plan are not met?

   Yes _____   No _____

   Describe:_____
   _____
   _____

7. Are the corrective actions the same as described in the plan?

   Yes _____   No _____

   Describe:_____
   _____
   _____

8. Who is responsible for verification that the required records are being properly maintained?

   Describe:_____
   _____
   _____

9. Did employees and managers demonstrate knowledge of the HACCP plan?

   Yes _____   No _____

   Describe:_____
   _____
   _____

10. What training has been provided to support the HACCP plan?

   Describe:_____
   _____
   _____

FORM (continued)

11. Describe examples of any documentation that the above training was accomplished.

Describe:_____

_____

_____

12. Are calibrations of equipment/thermometers performed as required by the plan?

Yes _____   No _____

Describe:_____

_____

_____

Additional Comments:

Person Interviewed:_____

Source: 2000 Food Code, Food and Drug Administration, U.S. Department of Health and Human Services, www.cfsan.fda.gov/~ear/rfi-x4-1.html.

FORM

HACCP Plan Verification Worksheet

Establishment Name:	Type of Facility:
Physical Address:	Person in Charge:

City:	State:	Zip:	County:

Inspection Time In:	Inspection Time Out:	Date:	Candidate's Name:

Agency:	Standard's Name:	Indicate Person Filling Out Form: (circle one) Candidate's Form/Standard's Form

Cold Holding Requirement for Jurisdiction: 41°F (5°C) _____ or 45°F (7°C) _____ or 41°F (5°C) and 45°F (7°C) combination: _____

Chart 2: HACCP Plan Verification Summary

HACCP Plan Verification Summary (circle YES or NO)

	Record #1	Record #2	Record #3
	Today's Date:	2nd Selected Date:	3rd Selected Date:
Required Monitoring Recorded [1]	YES / NO	YES / NO	YES / NO
Accurate and Consistent [2]	YES / NO	YES / NO	YES / NO
Corrective Action Documented [3]	YES / NO	YES / NO	YES / NO

Total # of record answers that are in Disagreement with the Standard = _____ (This box for Completion by Standard only.)

1. Have there been any changes to the food establishment menu?

   Yes _____   No _____

   Describe: _____

   _____

The use of a HACCP plan by a food establishment can be verified through a review of food establishment records and investigating the following information:

1. Does the food establishment's HACCP documentation indicate that required monitoring was recorded on the three selected dates? A "YES" answer would indicate that all required monitoring was documented. If any required monitoring was not documented, a "NO" answer would be circled in this section.

2. Does the food establishment's HACCP documentation for the selected dates appear accurate and consistent with other observations? A "YES" answer would indicate that the record appears accurate and consistent. A "NO" answer would indicate that there is inaccurate or inconsistent HACCP documentation.

3. Was corrective action documented in accordance with the HACCP plan when CLs were not met on each of the three selected dates? A "YES" answer would indicate that corrective action was documented for each CL not met for each of the three selected dates. A "YES" can also mean that no corrective action was needed. A "NO" answer would indicate any missing or inaccurate documentation of corrective action.

Here is a quick review of verification activities:

- ★ Review hazard analysis and CCPs.
- ★ Check critical control point records.
- ★ Validate critical limits.
- ★ Understand why foods haven't reached their critical limits.
- ★ Review monitoring records and frequency of activity.
- ★ Observe employees performing tasks, especially monitoring CCPs.
- ★ Check equipment temperatures.
- ★ Confirm all equipment is working properly.
- ★ Determine causes for equipment failure and procedure failure.
- ★ Ensure third-party or in-house validation is complete.

As a foodservice employee, realize that a new menu or a concept change will require verification and changes to your HACCP plan. **Verification is important because it reviews every Star Point and confirms that your HACCP plan is working.**

STAR KNOWLEDGE EXERCISE: VERIFICATION SCENARIOS

As the person-in-charge, how would you handle the following scenarios? Using the space provided, complete the scenario with your company procedures or best leadership practice to achieve successful verification. Then justify and explain why you took that course of action.

Scenario	What Do You Do? WHY?
1. As a manager performing verification, you notice that the cold-holding chart is perfect with every cold food product exactly at 41°F (5°C).	
2. During verification, you realize that the critical control point is not being met for cooking hamburgers.	

Scenario	What Do You Do? WHY?
3. A memo has been posted on the bulletin board: "Attention all Employees: HACCP Verification/Audit/Inspection is from the 10th to the 13th and everyone is expected to perform their very best, no matter what!" You're the person that arrives on the 10th to do the HACCP verification.	
4. After reviewing the first three pages of records, you notice that the same employee completed the records with the same pen for all 4 days, and then you look at the employee schedule and realize that employee was on vacation during the week the records were completed.	
5. Using the ice-point method of calibration, you check the thermometers used for monitoring and find the there are five thermometers being used and the temperatures at ice point are as follows: **a.** 33°F (0.6°C) **b.** 31°F (–0.6°C) **c.** 29°F (–1.7°C) **d.** 35°F (1.7°C) **e.** 35°F (1.7°C)	
6. You are observing an employee prepare and cook fresh chicken wings according to the standard operating procedure for frozen chicken wings.	
7. During verification, you observe an employee reheating meatballs for the third time in 3 hours.	

Scenario	What Do You Do? WHY?
8. During your verification visit a customer returns a chicken breast that is bloody in the center.	
9. A different customer brings back another bloody chicken sandwich and hands it to an employee. Then that same employee looks at you for direction.	
10. The microbiological testing results have come back positive for E. coli in the ground beef.	

PRINCIPLE 7: RECORD KEEPING AND DOCUMENTATION

To achieve the final HACCP principle, you must keep all of the documents created while establishing your HACCP plan. HACCP Principle 7 involves maintaining all the paperwork, documents, logs, and so on that you used to achieve active managerial control and protect the flow of food and the food safety management system. It is critical when documenting HACCP to get enough information to ensure your plan is effective. All records must be accurate and legible. As the manager or supervisor, it is your responsibility to ensure correct and accurate record keeping. The employees on your team will follow your example. For instance, never scribble or use correction tape or liquid because it may look like a cover-up. For errors, always draw a line through the mistake and initial it.

To be effective, your record-keeping plans should include the following:

Prerequisite Records:

★ Company Organization Chart

★ Employee Health Records* (FDA Forms: 1-A, 1-B, 1-C, 1-D)

★ Employee Training Log

- ★ Employee Food Defense Awareness*
- ★ Schedules
- ★ Standardized Recipes*
- ★ Equipment Monitoring Logs*
- ★ Equipment Maintenance Log
- ★ List of Approved Chemicals
- ★ Master Cleaning Checklist
- ★ Sanitizer Checklist*
- ★ Pest Control Documentation
- ★ Food Establishment Inspection Report*
- ★ SOPs*

Purchasing Records:

- ★ Supplier Certification Records
- ★ Processor Audit Record
- ★ Invoices
- ★ Shellstock Tags—mandatory records that must be filed in chronological order for at least 90 days

Receiving Records:

- ★ Receiving Log*
- ★ Receiving Reject Form*

Storing Records:

- ★ Storage Temperature Logs
- ★ Freezer Log
- ★ Refrigeration/Cooler Log*

Preparing Records:

- ★ Preparation Log
- ★ Grinding Log (Lot #'s)

Cooking Records:

- ★ Time-Temperature Logs*
- ★ Time-Temperature Graphs

Holding Records:

- ★ Cold-Holding Food Logs
- ★ Hot-Holding Food Logs
- ★ Cold-Holding Equipment*
- ★ Hot-Holding Equipment

Cooling Records:

- ★ Cooling Logs*

Reheating Records:

- ★ Reheating Logs*

Serving Records:

- ★ Serving Line Temperature Log
- ★ Cold-Serving Equipment*
- ★ Hot-Serving Equipment

TRAINING SCHEDULE

Food Safety SOPs

Temperature Log

PRINCIPLE 7

HACCP:

★ HACCP Plan Form*

Principle 1:

★ Hazard Analysis Work Sheet*

★ Flowcharts

★ Specifications of the Food Products

Principle 2:

★ Process 1—Food Preparation with No-Cook Step Chart*

★ Process 2—Preparation for Same-Day Service*

★ Process 3—Complex Food Preparation*

Principle 3:

★ Critical Limit Chart

★ Shelf Life Chart

Principle 4:

★ Calibrated Thermometers*

★ Monitoring Procedures

★ Monitoring Report*

Principle 5:

★ Corrective Actions Records

★ Discard Log (Waste Chart/Shrink Log)*

Principle 6:

★ Verification Records

Principle 7:

★ Foodborne Illness Investigation*

★ Food Safety Checklist*

★ Checklists

Note: Items marked with * have samples included.

This is a big list and at first glance, it is very daunting. But with the proper forms and training, the use of such documentation will become second nature to you and your employees. If you don't include such documentation in your HACCP plan, it will most likely fail.

Following are a sample of the types of forms used for record keeping.

■ SAMPLE RECORD-KEEPING CHARTS

FORM 1-A	**Conditional Employee and Food Employee Interview**
	Preventing Transmission of Diseases through Food by Infected Food Employees or Conditional Employees with Emphasis on illness due to Norovirus, *Salmonella* Typhi, *Shigella* spp., Enterohemorrhagic (EHEC) or Shiga toxin-producing *Escherichia coli* (STEC), or hepatitis A Virus

The purpose of this interview is to inform conditional employees and food employees to advise the person in charge of past and current conditions described so that the person in charge can take appropriate steps to preclude the transmission of foodborne illness.

Conditional employee name (print) _____

Food employee name (print) _____

Address _____

Telephone   *Daytime:*_____   *Evening:* _____

Date _____

Are you suffering from any of the following symptoms? (Circle one)

		If YES, Date <u>of Onset</u>
Diarrhea?	YES / NO	_____
Vomiting?	YES / NO	_____
Jaundice?	YES / NO	_____
Sore throat with fever?	YES / NO	_____

Or

Infected cut or wound that is open and draining, or lesions containing pus on the hand, wrist, an exposed body part, or other body part and the cut, wound, or lesion not properly covered?   YES / NO

(Examples: *boils and infected wounds, however small*)

<u>In the Past:</u>

Have you ever been diagnosed as being ill with typhoid fever (*Salmonella* Typhi)   YES / NO

If you have, what was the date of the diagnosis? _____

If within the past 3 months, did you take antibiotics for *S. Typhi*?   YES / NO

    If so, how many days did you take the antibiotics? _____

    If you took antibiotics, did you finish the prescription? _____   YES / NO

<u>History of Exposure:</u>

1.   Have you been suspected of causing or have you been exposed to a confirmed foodborne disease outbreak recently?   YES / NO

           If YES, date of outbreak: _____

a.   If YES, what was the cause of the illness and did it meet the following criteria?

    Cause: _____

    i.   Norovirus (last exposure within the past 48 hours)   Date of illness outbreak _____

    ii.   *E. coli* O157:H7 infection (last exposure within the past 3 days)   Date of illness outbreak _____

    iii.   Hepatitis A virus (last exposure within the past 30 days)   Date of illness outbreak _____

    iv.   Typhoid fever (last exposure within the past 14 days)   Date of illness outbreak _____

    v.   Shigellosis (last exposure within the past 3 days)   Date of illness outbreak _____

FORM 1-A (continued)

b. **If YES, did you:**
  i. **Consume food implicated in the outbreak?** _____
  ii. **Work in a food establishment that was the source of the outbreak?** _____
  iii. **Consume food at an event that was prepared by person who is ill?** _____

2. **Did you attend an event or work in a setting, recently where there was a confirmed disease outbreak?**   **YES / NO**

  If so, what was the cause of the confirmed disease outbreak? _____

  If the cause was one of the following five pathogens, did exposure to the pathogen meet the following criteria?

  a. **Norovirus (last exposure within the past 48 hours)**   **YES / NO**
  b. **E. coli O157:H7 (or other EHEC/STEC (last exposure within the past 3 days)**   **YES / NO**
  c. **Shigella spp. (last exposure within the past 3 days)**   **YES / NO**
  d. **S. Typhi (last exposure within the past 14 days)**   **YES / NO**
  e. **hepatitis A virus (last exposure within the past 30 days)**   **YES / NO**

  Do you live in the same household as a person diagnosed with Norovirus, Shigellosis, typhoid fever, hepatitis A, or illness due to E. coli O157:H7 or other EHEC/STEC?
                                  YES / NO Date of onset of illness _____

3. **Do you have a household member attending or working in a setting where there is a confirmed disease outbreak of Norovirus, typhoid fever, Shigellosis, EHEC/STEC infection, or hepatitis A?**
                                  YES / NO Date of onset of illness _____

  Name, Address, and Telephone Number of your Health Practitioner or doctor:
  Name _____
  Address _____
  Telephone – *Daytime:* _____ *Evening:* _____

Signature of Conditional Employee _____ **Date** _____

Signature of Food Employee _____ **Date** _____

Signature of Permit Holder or Representative _____ **Date** _____

FORM 1-B	**Conditional Employee or Food Employee Reporting Agreement**
	Preventing Transmission of Diseases through Food by Infected Conditional Employees or Food Employees with Emphasis on illness due to Norovirus, *Salmonella* Typhi, *Shigella* spp., Enterohemorrhagic (EHEC) or Shiga toxin-producing *Escherichia coli* (STEC), or hepatitis A Virus

The purpose of this agreement is to inform conditional employees or food employees of their responsibility to notify the person in charge when they experience any of the conditions listed so that the person in charge can take appropriate steps to preclude the transmission of foodborne illness.

I AGREE TO REPORT TO THE PERSON IN CHARGE:

Any Onset of the Following Symptoms, Either While at Work or Outside of Work, Including the Date of Onset:

    1. Diarrhea
    2. Vomiting
    3. Jaundice
    4. Sore throat with fever
    5. Infected cuts or wounds, or lesions containing pus on the hand, wrist , an exposed body part, or other body part and the cuts, wounds, or lesions are not properly covered (*such as boils and infected wounds, however small*)

Future Medical Diagnosis:

Whenever diagnosed as being ill with Norovirus, typhoid fever (*Salmonella* Typhi), shigellosis (*Shigella* spp. infection), *Escherichia* coli O157:H7 or other EHEC/STEC infection, or hepatitis A (hepatitis A virus infection)

Future Exposure to Foodborne Pathogens:

1. Exposure to or suspicion of causing any confirmed disease outbreak of Norovirus, typhoid fever, shigellosis, *E.* coli O157:H7 or other EHEC/STEC **infection, or hepatitis A.**
2. A household member diagnosed with Norovirus, typhoid fever, shigellosis, illness due to EHEC/STEC, **or hepatitis A.**
3. A household member attending or working in a setting experiencing a confirmed disease outbreak of Norovirus, typhoid fever, shigellosis, *E.* coli O157:H7 or other EHEC/STEC **infection, or hepatitis A.**

I have read (or had explained to me) and understand the requirements concerning my responsibilities under the **Food Code** and this agreement to comply with:

1. Reporting requirements specified above involving symptoms, diagnoses, and exposure specified;
2. Work restrictions or exclusions that are imposed upon me; and
3. Good hygienic practices.

I understand that failure to comply with the terms of this agreement could lead to action by the food establishment or the food regulatory authority that may jeopardize my employment and may involve legal action against me.

Conditional Employee Name (please print) _____

Signature of Conditional Employee _____ **Date** _____

Food Employee Name (please print) _____

Signature of Food Employee _____ **Date** _____

Signature of Permit Holder or Representative _____ **Date** _____

FORM 1-C   **Conditional Employee or Food Employee Medical Referral**

Preventing Transmission of Diseases through Food by Infected Food Employees with Emphasis on Illness due to Norovirus, Typhoid fever (*Salmonella* Typhi), Shigellosis (*Shigella* spp.), *Escherichia coli* O157:H7 or other Enterohemorrhagic (EHEC) or Shiga toxin-producing *Escherichia* coli (STEC), and hepatitis A Virus

The **Food Code** specifies, under *Part 2-2 Employee Health Subpart 2-201 Disease or Medical Condition,* that Conditional Employees and Food Employees obtain medical clearance from a health practitioner licensed to practice medicine, unless the Food Employees have complied with the provisions specified as an alternative to providing medical documentation, whenever the individual:

1.   Is chronically suffering from a symptom such as **diarrhea;** *or*
2.   Has a **current illness** involving Norovirus, typhoid fever (*Salmonella* **Typhi**), shigellosis (*Shigella* spp.) *E. coli* **O157:H7** infection (or other EHEC/STEC), or hepatitis A virus (hepatitis A), *or*
3.   Reports *past illness* involving typhoid fever (*S. Typhi*) within the past three months (while salmonellosis is fairly common in U.S., typhoid fever, caused by infection with *S. Typhi*, is rare).

Conditional employee being referred: (Name, please print) _____

Food Employee being referred: (Name, please print) _____

4.   Is the employee assigned to a food establishment that serves a population that meets the Food Code definition of a **highly susceptible population** such as a day care center with preschool age children, a hospital kitchen with immunocompromised persons, or an assisted living facility or nursing home with older adults?
YES ☐   NO ☐

Reason for Medical Referral: The reason for this referral is checked below:
☐   Is chronically suffering from vomiting or diarrhea; or (specify) _____
☐   Diagnosed or suspected Norovirus, typhoid fever, shigellosis, *E. coli* O157:H7 (or other EHEC/STEC) infection, or hepatitis A. (Specify) _____
☐   Reported past illness from typhoid fever within the past 3 months. (Date of illness) _____
☐   Other medical condition of concern per the following description: _____

Health Practitioner's Conclusion: (Circle the appropriate one; refer to reverse side of form)
☐   Food employee is free of **Norovirus** infection, typhoid fever (*S. Typhi* infection), *Shigella* spp. infection, *E. coli* O157:H7 (or other **EHEC/STEC** infection), or **hepatitis A** virus infection, and may work as a food employee without restrictions.
☐   Food employee is an asymptomatic shedder of *E.* coli O157:H7 (or other **EHEC/STEC**), *Shigella* spp., or Norovirus, and is restricted from working with exposed food; clean equipment, utensils, and linens; and unwrapped single-service and single-use articles in food establishments that do not serve highly susceptible populations.
☐   Food employee is not ill but continues as an asymptomatic shedder of *E. coli* O157:H7 (or other **EHEC/STEC**), *Shigella* spp. and should be excluded from food establishments that serve highly susceptible populations such as those who are preschool age, immunocompromised, or older adults and in a facility that provides preschool custodial care, health care, or assisted living.
☐   Food employee is an asymptomatic shedder of **hepatitis A** virus and should be excluded from working in a food establishment until medically cleared.
☐   Food employee is an asymptomatic shedder of **Norovirus** and should be excluded from working in a food establishment until medically cleared, or for at least 24 hours from the date of the diagnosis.
☐   Food employee is suffering from Norovirus, typhoid fever, shigellosis, *E. coli* O157:H7 (or other **EHEC/STEC** infection), or **hepatitis A** and should be excluded from working in a food establishment.

FORM 1-C (continued)

COMMENTS: (In accordance with Title I of the Americans with Disabilities Act (ADA) and to provide only the information necessary to assist the food establishment operator in preventing foodborne disease transmission, please confine comments to explaining your conclusion and estimating when the employee may be reinstated.)

_____

_____

_____

_____

_____

_____

_____

Signature of Health Practitioner _____ **Date** _____

Paraphrased from the FDA Food Code for Health Practitioner's Reference

From Subparagraph 2-201.11(A)(2)　　　Organisms of Concern:

Any foodborne pathogen, with special emphasis on these 5 organisms:
1. **Norovirus** 2. *S.* **Typhi** 3. *Shigella* spp. 4. *E. coli* O157:H7 (or other EHEC/STEC) 5. **Hepatitis A** virus

From Subparagraph 2-201.11(A)(1)　　　Symptoms:

Have any of the following symptoms:
　　Diarrhea　　　　**Vomiting**　　　　**Jaundice**　　　　**Sore throat with fever**

From Subparagraph 2-201.11(A)(4)-(5)　　　Conditions of Exposure of Concern:

(1) Suspected of causing a foodborne outbreak or being exposed to an outbreak caused by 1 of the 5 organisms above, at an event such as a family meal, church supper, or festival because the person:
　　Prepared or consumed an implicated food; or
　　Consumed food prepared by a person who is infected or ill with the organism that caused the outbreak or who is suspected of being a carrier;
(2) Lives with, and has knowledge about, a person who is diagnosed with illness caused by 1 of the 5 organisms; or
(3) Lives with, and has knowledge about, a person who works where there is an outbreak caused by 1 of the 5 organisms.

From Subparagraph 2-201.12　　　Exclusion and Restriction:

Decisions to exclude or restrict a food employee are made considering the available evidence about the person's role in actual or potential foodborne illness transmission. Evidence includes:

　　Symptoms　　　**Diagnosis**　　　**Past illnesses**　　　**Stool/blood tests**

In facilities serving highly susceptible populations such as day care centers and health care facilities, a person for whom there is evidence of foodborne illness is almost always <u>excluded</u> from the food establishment.

In other establishments such as restaurants and retail food stored, that offer food to typically healthy consumers, a person might only be <u>restricted</u> from certain duties, based on the evidence of foodborne illness.

Exclusion from any food establishment is required when the person is:
- Exhibiting or reporting diarrhea or vomiting;
- Diagnosed with illness caused by *S.* Typhi; or
- Jaundiced within the last 7 days.

For *Shigella* spp. or *Escherichia coli* O157:H7 or other EHEC/STEC infections, the person's stools must be negative for 2 consecutive cultures taken no earlier than 48 hours after antibiotics are discontinued, and at least 24 hours apart or the infected individual must have resolution of symptoms for more than 7 days or at least 7 days have passed since the employee was diagnosed.

FORM 1-D	**Application for Bare Hand Contact Procedure** (As specified in Food Code ¶ 3-301.11(D))

Please type or print legibly using black or blue ink

1. **Establishment Name:** _____

2. **Establishment Address:** _____

3. **Responsible Person:** _____ **Phone:** _____

Legal Representative   Business

4. **List Procedure and Specific Ready-To-Eat-Foods** to be considered for use of bare hand contact with ready-to-eat foods:
   _____
   _____
   _____

5. **Handwashing Facilities**:

   (a) There is a handwashing sink located immediately adjacent to the posted bare hand contact procedure and the hand sink is maintained in accordance with provisions of the Code. (§ 5-205.11, § 6-301.11, § 6-301.12, § 6-301.14) □ YES □ NO (Include diagram, photo or other information)

   (b) All toilet rooms have one or more handwashing sinks in, or immediately adjacent to them, and the sinks are equipped and maintained in accordance with provisions of the Code. (§ 5-205.11, § 6-301.11, § 6-301.12, § 6-301.14) □ YES □ NO

6. **Employee Health Policy:** The written employee health policy must be attached to this form along with documentation that food employees and conditional employees acknowledge their responsibilities. (§ 2-201.11, § 2-201.12, § 2-201.13)

7. **Employee Training**: Provide documentation that food employees have received training in:

   - The risks of contacting the specific ready-to-eat foods with bare hands
   - Personal health and activities as they relate to diseases that are transmissible through food.
   - Proper handwashing procedures to include how, when, where to wash, & fingernail maintenance. (§ 2-301.12, § 2-301.14, § 2-301.15, § 2-302.11)
   - Prohibition of jewelry. (§ 2-303.11)
   - Good hygienic practices. (§ 2-401.11, § 2-401.12)

8. **Documentation of Handwashing Practices:** Provide documentation that food employees are following proper handwashing procedures prior to food preparation and other procedures as necessary to prevent cross-contamination during all hours of operation when the specific ready-to-eat foods are prepared or touched with bare hands.

9. **Documentation of Additional Control Measures:** Provide documentation to demonstrate that food employees are utilizing two or more of the following control measures when contacting ready-to-eat foods with bare hands:

   - Vaccination against hepatitis A for food employees including initial booster shots or documented medical evidence that a food employee has had a previous illness from hepatitis A virus;
   - Double handwashing;
   - Use of nailbrushes;
   - Use of hand antiseptic after handwashing;
   - Incentive programs such as paid leave encouraging food employees not to work when they are ill; or
   - Other control measures approved by the regulatory authority.

Statement of Compliance:

I certify all of the following: All food employees are individually trained in the risks of contacting ready-to-eat foods with bare hands, personal health and activities as they relate to diseases that are transmissible through food, proper handwashing procedures, prohibition of jewelry, and good hygienic practices. A record of this training is kept on site. I understand that bare hand contact with ready-to-eat food is prohibited except for those items listed in section four (4) above. A handwashing sink is located immediately adjacent to the posted bare hand contact procedure. All handwashing sinks are maintained with hot water, soap, and drying devices. I understand that documentation is needed for handwashing practices and additional control measures. I understand that records to document handwashing are kept current and kept on site.

SIGNATURE:_____ DATE _____

(Signature of legal representative of the facility listed above)

Regulatory Authority (RA) Use Only:
Permit Number: _____
File Review Conducted on History of Handwashing Compliance: □ Yes □ No
Site Visit Conducted □ Yes □ No Comments: _____
□ Approved: Effective Date: _____ RA name _____
□ Not Approved: Reason for Denial: _____

Form: Employee Food Defense Awareness

Establishment's Name:

Employee Name: _____ **Employee ID#:** _____

EMPLOYEE AWARENESS SOP: Employee Initials

★ Be a responsible employee. Communicate any potential food defense issues to your manager. _____

★ Be aware of your surroundings and pay close attention to customers and employees who are acting suspiciously. _____

★ Limit the amount of personal items you bring into your work establishment. _____

★ Be aware of who is working at a given time and where (in what area) they are supposed to be working. _____

★ Periodically monitor the salad bar and food displays. _____

★ Make sure labeled chemicals are in a designated storage area. _____

★ Make sure you and your coworkers are following company guidelines. If you have any questions or feel as though company guidelines are not being followed, please ask your manager to assist you. _____

★ Take all threats seriously, even if it is a fellow coworker blowing off steam about your manager and what he or she wants to do to get back at your manager or your company; or if he or she is angry and wants to harm the manager, the customer, or the business. _____

★ If the back door is supposed to be locked and secure, **make sure it is!** _____

★ If you use a food product every day and it is supposed to be blue but today it is green, stop using the product and notify your manager. _____

★ If you know an employee is no longer with your company and this person enters an "Employees Only" area, notify your manager immediately. _____

★ Cooperate in all investigations. _____

★ Do not talk to the media; refer all questions to your corporate office. _____

★ If you are aware of a hoax, notify your manager immediately. _____

CUSTOMER AWARENESS SOP:

★ Be aware of any unattended bags or briefcases customers bring into your operation. _____

★ If a customer walks into an "Employee Only" area of your operation, ask the customer politely if he or she needs help, then notify a member of management. _____

VENDOR AWARENESS SOP:

★ Check the identification of any vendor or service person that enters restricted areas of your operation and do not leave him or her unattended. _____

★ Monitor all products received and look for any signs of tampering. _____

★ When a vendor is making a delivery, never accept more items than what is listed on your invoice. If the vendor attempts to give you more items than what is listed, notify your manager. _____

★ When receiving deliveries:
Step 1. Always ask for identification. _____
Step 2. Stay with the delivery person. _____
Step 3. Do not allow the person to roam freely throughout your operation. _____

FACILITY AWARENESS SOP:

★ Report all equipment, maintenance, and security issues to your manager.

★ Document any equipment, maintenance, and security issues.

★ Be aware of the inside and outside of your facility, including the dumpster area, and report anything out of the ordinary.

ESTABLISHMENT-SPECIFIC FOOD DEFENSE (OPTIONAL):

Employee Signature: Date: Person in Charge Signature:

Standardized Recipes

A standardized USDA recipe checklist:

- ☐ Name of recipe
- ☐ Ingredients list
- ☐ Weight and measure
- ☐ Preparation directions
- ☐ Serving directions

- ☐ Yield
- ☐ Portion size information
- ☐ Variations
- ☐ Nutrients per serving
- ☐ Pan size if appropriate

Chili con Carne with Beans

Meat/Meat Alternate-Vegetable   Main Dishes   D-20

Ingredients	50 servings		100 Servings		Directions
	Weight	Measure	Weight	Measure	
Raw ground beef (no more than 20% fat)	7 lb		14 lb		1. Brown ground beef. Drain. Continue immediately.
*Fresh onions, chopped *or* Dehydrated onions	14 oz *or* 2-1/2 oz	2-1/2 cups *or* 1-3/4 cup	1 lb 12 oz *or* 5 oz	1 qt 2/3 cup *or* 2-1/2 cups	2. Add onions, granulated garlic, green pepper (optional), pepper, chili powder, paprika, onion powder, and ground cumin. Cook for 5 minutes.
Granulated garlic		1 Tbsp 1-1/2 tsp		3 Tbsp	
*Fresh green pepper chopped (optional)	8 oz	1-1/2 cups, 2 Tbsp	1 lb	3-1/4 cups	
Ground black or white pepper		2 tsp		1 Tbsp 1 tsp	
Chili powder		3 Tbsp		1/4 cup 2 Tbsp	
Paprika		1 Tbsp		2 Tbsp	
Onion powder		1 Tbsp		2 Tbsp	
Ground cumin	1 oz	1/4 cup	2 oz	1/2 cup	
Canned diced tomatoes, with juice	3 lb 3 oz	1 qt 2-1/4 cups (1/2 No. 10 can)	6 lb 6 oz	3 qt 1/2 cup (1 No. 10 can)	3. Stir in tomatoes, water, and tomato paste; mix well. Bring to boil. Reduce heat. Cover. Simmer slowly, stirring occasionally until thickened, about 40 minutes.
Water		2 qt 1 cup		1 gal 2 cups	

FORM (continued)

Chili con Carne with Beans

| Meat/Meat Alternate-Vegetable | | | | Main Dishes | D-20 |

Ingredients	50 servings		100 Servings		Directions
	Weight	Measure	Weight	Measure	
Canned tomato paste	1 lb 12 oz	3 cups 2 Tbsp (1/4 No. 10 can)	3 lb 8 oz	1 qt 2-1/4 cups (1/2 No. 10 can)	
Canned pinto or kidney beans, drained *or* *Dry pinto or kidney beans, cooked (see Special Tip)	3 lb 6 oz *or* 2 lb 4 oz	1 qt 3-1/2 cups (1/2 No. 10 can) *or* 1 qt 2 cups	6 lb 12 oz *or* 4 lb 8 oz	3 qt 3 cups (1 No. 10 can) *or* 3 qt	4. Stir in beans. Cover and simmer. Stir occasionally. CCP: Heat to 155°F (68.3°C) or higher for 15 seconds. *or* If using previously cooked and chilled beans: CCP: Heat to 165°F (73.9°C) or higher for at least 15 seconds.
					5. Pour into serving pans.
					6. CCP: Hold for hot service at 135°F (57.2°C or higher). Portion with 4 oz ladle (1/2 cup).
Reduced fat Cheddar cheese, shredded (optional)	1 lb 8 oz	1 qt 2 cups	3 lb	3 qt	7. Garnish with cheese (optional).

FORM (continued)

Chili con Carne with Beans

Meat/Meat Alternate-Vegetable   Main Dishes   D-20

Comments:
*See Marketing Guide

Marketing Guide for Selected Items		
Food as Purchased for	**50 Servings**	**100 Servings**
Mature onions	1 lb	2 lb
Green peppers	11 oz	1 lb 6 oz
Dry pinto beans, dry	1 lb	2 lb
or	*or*	*or*
Dry kidney beans	1 lb	2 lb

Serving:

1/2 cup (4 oz ladle) provides 2 oz equivalent meat/meat alternate and 3/8 cup of vegetable.

Yield:

50 Servings: about 16 lb 4 oz

100 Servings: about 32 lb 8 oz

Tested 2004

Volume:

50 Servings: about 1 gallons 2-1/4 quarts

100 Servings: about 3 gallons 2 cups

Special Tip:

Soaking Beans

Overnight method: Add 1-3/4 qt cold water to every 1 lb of dry beans. Cover and refrigerate overnight. Discard the water. Proceed with recipe.

Quick-soak method: Boil 1-3/4 qt of water for each 1 lb of dry beans. Add beans and boil for 2 minutes. Remove from heat and allow to soak for 1 hour. Discard the water. Proceed with recipe.

Cooking Beans

Once the beans have been soaked, add 1/2 tsp salt for every lb of dry beans. Boil gently with lid tilted until tender, about 2 hours.

Use hot beans immediately.
CCP: Hold for hot service at 135°F (57.2°C).
or
Or, chill for later use.
CCP: Cool to 70°F (21.1°C) within 2 hours and to 41°F (5°C) or lower within an additonal 4 hours.

1 lb dry pinto beans = about 2-3/8 cups dry or 5-1/4 cups cooked beans.
1 lb dry kidney beans = about 2-1/2 cups dry or 6-1/4 cups cooked beans.

Variation:

A. Chili con Carne without Beans

50 servings: In step 1, use 8 lb 10 oz raw ground beef. Continue with steps 2 and 3. In step 4, omit pinto beans or kidney beans. Continue with step 5–7.

100 servings: In step 1, use 17 lb 4 oz raw ground beef. Continue with steps 2 and 3. In step 4, omit pinto beans or kidney beans. Continue with step 5–7.

Chili con Carne with Beans

Meat/Meat Alternate-Vegetable   Main Dishes   D-20

Nutrients Per Serving

Calories	180	Saturated Fat	3.57 g	Iron	2.71 mg
Protein	15.44 g	Cholesterol	42 mg	Calcium	46 mg
Carbohydrate	10.68 g	Vitamin A	813 IU	Sodium	204 mg
Total Fat	8.58 g	Vitamin C	14.5 mg	Dietary Fiber	2.5 g

FORM

Tomato Sauce - Canned

For use in the USDA Household Commodity Food Distribution Programs
03/14/03

Product Description

Canned Tomato Sauce is U.S. Grade A. It may be lightly seasoned with salt and spices, and may contain nutritive sweetening ingredients, vinegar, onion, garlic, or other vegetable flavoring ingredients.

Pack/Yield

Tomato Sauce is packed in a 15.5-ounce can, which yields about 3 1/2-cup servings

Storage

- Store unopened cans in a cool, dry place off the floor.
- Avoid freezing or exposure to direct sunlight. Sudden changes in temperature shorten shelf life and speed deterioration.
- Store opened tomato sauce in a tightly covered nonmetallic container and refrigerate. Use within 2 to 4 days.

Preparation

Canned tomato sauce is ready to use. Preparation depends on final use and may be part of recipe instructions.

Uses

- Use tomato sauce as an ingredient in sauces, stews, casseroles, pizza, and soups.
- Use as topping for cooked pasta, or as an ingredient in other Italian style dishes.
- Serve warm as a dipping sauce for breadsticks.

Nutrition Information

- **Tomato Sauce** is an excellent source of Vitamin A and Vitamin C.
- 1/2 cup of tomato sauce provides 1 serving from the **Vegetable Group** of the **Food Guide Pyramid.**

(See recipes on next page)

Nutrition Facts
Serving size 1/2 cup (122g) Tomato Sauce

Amount per Serving			
Calories	45	Fat Calories	0
		% Daily Value*	
Total Fat	0g		0%
Saturated Fat	0g		0%
Cholesterol	0mg		0%
Sodium	740mg		31%
Total Carbohydrate	9g		3%
Dietary Fiber	2g		8%
Protein	2g		
Vitamin A	25%	Vitamin C	25%
Calcium	0%	Iron	6%

*Percent Daily Values are based on a 2,000 calorie diet.

FORM

SLOPPY JOE MEATBALLS (Makes 6 servings)

1 egg, beaten

1/4 cup fine dry bread crumbs

1 medium onion, finely chopped (1/2 cup)

1/4 teaspoon dried oregano, crushed

1 pound lean ground beef or ground bison

1/2 cup green bell pepper, chopped

1 tablespoon vegetable oil

1 (15.5 ounce) can tomato sauce

2 tablespoons brown sugar

1 tablespoon prepared mustard

1 teaspoon chili powder

1/4 teaspoon ground black pepper

1/4 teaspoon garlic salt

Dashed bottled hot pepper sauce (optional)

Recipe provided by bhg.com (Better Homes & Gardens)

1. Heat oven to 350°F (176.7°C). Combine egg, bread crumbs, 1/4 cup of the onion, and the oregano in a large mixing bowl. Add the ground meat and mix well.

2. Shape into 42 meatballs about 3/4-inch in diameter. Arrange in a single layer in a 15 × 10 × 1-inch baking pan. Bake for 12 to 15 minutes, or until internal temperature is at least 155°F (68.3°C). Drain well.

3. Meanwhile, cook remaining 1/4 cup onion and the green pepper in hot oil in a large saucepan until vegetables are tender. Stir in tomato sauce, brown sugar, mustard, chili powder, black pepper, garlic salt, and if desired, hot pepper sauce. Bring to boiling.

4. Add meatballs to sauce. Reduce heat and simmer, uncovered, for 5 minutes.

Nutrition information for each serving of Sloppy Joe Meatballs:

Calories	310	Cholesterol	110 mg	Sugar	6 g	Calcium	44 mg
Calories from fat	160	Sodium	620 mg	Protein	25 g	Iron	3 mg
Total fat	18 g	Total Carbohydrate	14 g	Vitamin A	106 RE		
Saturated fat	6 g	Dietary fiber	2 g	Vitamin C	20 mg		

These recipes, presented to you by USDA, have not been tested or standardized.

U.S. Department of Agriculture * Food and Nutrition Service

USDA prohibits discrimination in all its programs and activities on the basis of race, color, national origin, sex, age, or disability. To file a complaint of discrimination, write USDA, Director, Office of Civil Rights, Room 326-W Whitten Building, 14th & Independence Avenue, SW, Washington, DC 20250-9410 or call (202) 720-5964 (voice and TDD). USDA is an equal opportunity provider and employer.

FORM

HEARTY BUFFET CHILI (Makes 6 servings)

1 pound lean ground beef or ground bison

1/3 cup chopped onion

1 clove garlic, minced

1 (15.5 ounce) can tomatoes, cut up

1 (15.5 ounce) can tomato sauce

2 teaspoons chili powder

1 teaspoon salt

1/2 teaspoon crushed red pepper

1/4 teaspoon black pepper

1 (15.5 ounce) can kidney beans

1 to 2 cups cooked rice

1/2 cup (4 ounces) shredded Monterey Jack
or cheddar cheese (optional)

1/2 cup chopped onion (optional)

1/2 cup dairy sour cream (optional)

Recipe provided by bhg.com (Better Homes & Gardens)

1. In a large saucepan, cook beef, the 1/3 cup onion and the garlic until beef is brown and onion is tender. Drain off fat.

2. Stir tomato sauce, undrained tomatoes, chili powder, salt, crushed red pepper and black pepper into browned meat in saucepan. Bring to boiling. Reduce heat. Cover and simmer for 1 hour.

3. Add beans with chili sauce to saucepan. Cover; simmer for 30 minutes.

4. In small individual bowls, set out remaining ingredients. For each serving, place some of the cooked rice in a soup bowl. Top with chili mixture and cheese, then some of the 1/2 cup onion and the sour cream, if you like.

Nutrition information for each serving of Hearty Buffet Chili

Calories	350	Cholesterol	76 mg	Sugar	6 g	Calcium	64 mg
Calories from fat	130	Sodium	970 mg	Protein	28 g	Iron	4 mg
Total fat	14 g	Total Carbohydrates	28 g	Vitamin A	117 RE		
Saturated fat	5 g	Dietary fiber	7 g	Vitamin C	21 mg		

These recipes, presented to you by USDA, have not been tested or standardized.

U.S. Department of Agriculture * Food and Nutrition Service

USDA prohibits discrimination in all its programs and activities on the basis of race, color, national origin, sex, age, or disability. To file a complaint of discrimination, write USDA, Director, Office of Civil Rights, Room 326-W Whitten Building, 14th & Independence Avenue, SW, Washington, DC 20250-9410 or call (202) 720-5964 (voice and TDD). USDA is an equal opportunity provider and employer.

Equipment Monitoring Chart

Equipment Monitoring Chart

Week Ending:

Menu Items:	Saturday		Sunday		Monday		Tuesday		Wednes.		Thursday		Friday	
	a.m.	p.m.	a.m.	p.m.	a.m.	p.m.	a.m.	p.m.	a.m.	p.m.	a.m.	p.m.	a.m.	p.m.
Freezer #1														
#2														
Freezer #1														
#2														
Products														
#1														
#2														
#3														
Deliveries														
#1														
#2														
#3														
Sanitizer (ppm)														
#1														
#2														
#3														
#4														

Sanitizer Checklist

Sanitizer Checklist

Date: **Taken by:**

		Comments
Sanitizer in Use	Y/N	_____
Quat 200 ppm	ppm	_____
Chlorine 100–200 ppm	ppm	_____
Wiping cloths sanitized	Y/N	_____
Equipment Sanitized		
Cutting boards	Y/N	_____
Knives	Y/N	_____
Slicers	Y/N	_____
Work surfaces	Y/N	_____
Prep sinks	Y/N	_____
Pocket thermometers	Y/N	_____
Thermometers sanitized each use	Y/N	_____

Food Establishment Inspection Report

FORM 3-A

Food Establishment Inspection Report

Page _____ of _____

As Governed by State Code Section			No. of Risk Factor/Intervention Violations		Date _____
			No. of Repeat Risk Factor/Intervention Violations		Time In _____
			Score (optional)		Time Out _____
Establishment	Address		City/State	Zip Code	Telephone
License/Permit #	Permit Holder		Purpose of Inspection	Est. Type	Risk Category

FOODBORNE ILLNESS RISK FACTORS AND PUBLIC HEALTH INTERVENTIONS

Circle designated compliance status (IN, OUT, N/O, N/A) for each numbered item   Mark "X" in appropriate box for COS and R

IN=in compliance   OUT=not in compliance   N/O=not observed   N/A=not applicable   COS=corrected on-site during inspection   R=repeat violation

	Compliance Status		COS	R			Compliance Status		COS	R
	Supervision						**Potentially Hazardous Food (TCS food)**			
1	IN OUT	Person in charge present, demonstrates knowledge, and performs duties				16	IN OUT N/A N/O	Proper cooking time and temperatures		
	Employee Health					17	IN OUT N/A N/O	Proper reheating procedures for hot holding		
2	IN OUT	Management awareness; policy present				18	IN OUT N/A N/O	Proper cooling time and temperatures		
3	IN OUT	Proper use of reporting, restriction & exclusion				19	IN OUT N/A N/O	Proper hot holding temperatures		
	Good Hygienic Practices					20	IN OUT N/A	Proper cold holding temperatures		
4	IN OUT N/O	Proper eating, tasting, drinking, or tobacco use				21	IN OUT N/A	Proper date marking and disposition		
5	IN OUT N/O	No discharge from eyes, nose, and mouth				22	IN OUT N/A N/O	Time as a public health control: procedures & records		
	Preventing Contamination by Hands						**Consumer Advisory**			
6	IN OUT N/O	Hands clean and properly washed				23	IN OUT N/A	Consumer advisory provided for raw or undercooked foods		
7	IN OUT N/A N/O	No bare hand contact with ready-to-eat foods or approved alternate method properly followed					**Highly Susceptible Populations**			
8	IN OUT	Adequate handwashing facilities supplied & accessible				24	IN OUT N/A	Pasteurized foods used; prohibited foods not offered		
	Approved Source						**Chemical**			
9	IN OUT	Food obtained from approved source				25	IN OUT N/A	Food additives: approved and properly used		
10	IN OUT N/A N/O	Food received at proper temperature				26	IN OUT	Toxic substances properly identified, stored, used		
11	IN OUT	Food in good condition, safe, and unadulterated					**Conformance with Approved Procedures**			
12	IN OUT N/A N/O	Required records available: shellstock tags, parasite destruction				27	IN OUT N/A	Compliance with variance, specialized process, and HACCP plan		
	Protection from Contamination									
13	IN OUT N/A	Food separated and protected								
14	IN OUT N/A	Food-contact surfaces: cleaned & sanitized								
15	IN OUT	Proper disposition of returned, previously served, reconditioned, and unsafe food								

> **Risk factors** are food preparation practices and employees behaviors most commonly reported to the Centers for Disease Control and Prevention as contributing factors in foodborne illness outbreaks.
> **Public health interventions** are control measures to prevent foodborne illness or injury.

GOOD RETAIL PRACTICES

Good Retail Practices are preventative measures to control the introduction of pathogens, chemicals, and physical objects into foods.

Mark "X" in box if numbered item is **not** in compliance   Mark "X" in appropriate box for COS and/or R   COS=corrected on-site during inspection   R=repeat violation

		COS	R			COS	R
	Safe Food and Water					**Proper Use of Utensils**	
28	Pasteurized eggs used where required			41	In-use utensils: properly stored		
29	Water and ice from approved source			42	Utensils, equipment and linens: properly stored, dried, handled		
30	Variance obtained for specialized processing methods			43	Single-use/single-service articles: properly stored, used		
	Food Temperature Control			44	Gloves used properly		
31	Proper cooling methods used; adequate equipment for temperature control				**Utensils, Equipment and Vending**	
32	Plant food properly cooked for hot holding			45	Food and nonfood-contact surfaces cleanable, properly designed, constructed, and used		
33	Approved thawing methods used			46	Warewashing facilities: installed, maintained, used; test strips		
34	Thermometers provided and accurate			47	Nonfood-contact surfaces clean		
	Food Identification				**Physical Facilities**	
35	Food properly labeled; original container			48	Hot and cold water available; adequate pressure		
	Prevention of Food Contamination			49	Plumbing installed; proper backflow devices		
36	Insects, rodents, and animals not present			50	Sewage and waste water properly disposed		
37	Contamination prevented during food preparation, storage & display			51	Toilet facilities: properly constructed, supplied, cleaned		
38	Personal cleanliness			52	Garbage/refuse properly disposed; facilities maintained		
39	Wiping cloths: properly used and stored			53	Physical facilities installed, maintained, and clean		
40	Washing fruits and vegetables			54	Adequate ventilation and lighting; designated areas use		

Person in Charge (Signature)   Date:

Inspector (Signature)   Follow-up: YES   NO   (Circle one)   Follow-up Date:

As these sample forms indicate, record keeping provides us with a history that we are following standard operating procedures. By documenting these steps, we are proving that we are consistently preparing, serving, and selling safe food. The additional benefit for documenting these steps is a more efficient operation, better control of costs, and the security of knowing you have done everything possible to serve food safely.

Sample SOP for Receiving Procedures

SOP: Receiving Deliveries (Sample)

Purpose: To ensure that all food is received fresh and safe when it enters the foodservice operation, and to transfer food to proper storage as quickly as possible

Scope: This procedure applies to foodservice employees who handle, prepare, or serve food.

Key Words: Cross-contamination, temperatures, receiving, holding, frozen goods, delivery

Instructions:

1. Train foodservice employees who accept deliveries on proper receiving procedures.
2. Schedule deliveries to arrive at designated times during operational hours.
3. Post the delivery schedule, including the names of vendors, days and times of deliveries, and drivers' names.
4. Establish a rejection policy to ensure accurate, timely, consistent, and effective refusal and return of rejected goods.
5. Organize freezer and refrigeration space, loading docks, and storerooms before deliveries.
6. Gather product specification lists and purchase orders, temperature logs, calibrated thermometers, pens, flashlights, and use clean loading carts before deliveries.
7. Keep receiving area clean and well lighted.
8. Do not touch ready-to-eat foods with bare hands.
9. Determine whether foods will be marked with the date of arrival or the "use-by" date, and mark accordingly upon receipt.
10. Compare delivery invoice against products ordered and products delivered.
11. Transfer foods to their appropriate locations as quickly as possible.

Monitoring:

1. Inspect the delivery truck when it arrives to ensure that it is clean, free of putrid odors, and organized to prevent cross-contamination. Be sure refrigerated foods are delivered on a refrigerated truck.
2. Check the interior temperature of refrigerated trucks.
3. Confirm vendor name, day and time of delivery, as well as driver's identification before accepting delivery. If the driver's name is different than what is indicated on the delivery schedule, contact the vendor immediately.
4. Check frozen foods to ensure that they are all frozen solid and show no signs of thawing and refreezing, such as the presence of large ice crystals or liquids on the bottom of cartons.

5. Check the temperature of refrigerated foods.

   a. For fresh meat, fish, and poultry products, insert a calibrated, clean and sanitized thermometer into the center of the product to ensure a temperature of 41°F (5°C) or below. The temperature of milk should be 45°F (7.2°C) or below.

   b. For packaged products, insert a food thermometer between two packages, being careful not to puncture the wrapper. If the temperature exceeds 41°F (5°C), it may be necessary to take the internal temperature before accepting the product.

   c. For eggs, the interior temperature of the truck should be 45°F (7.2°C) or below.

6. Check dates of milk, eggs, and other perishable goods to ensure safety and quality.

7. Check the integrity of food packaging.

8. Check the cleanliness of crates and other shipping containers before accepting products. Reject foods that are shipped in dirty crates.

Corrective Action:

1. Reject the following:

   a. Frozen foods with signs of previous thawing

   b. Cans that have signs of deterioration—swollen sides or ends, flawed seals or seams, dents, or rust

   c. Punctured packages

   d. Expired foods

   e. Foods that are out of the safe temperature zone or deemed unacceptable by the established rejection policy

Verification and Record Keeping:

Record temperature and corrective action on the delivery invoice or on the receiving log. The foodservice manager will verify that foodservice employees are receiving products using the proper procedure by visually monitoring receiving practices during the shift and reviewing the receiving log at the close of each day. Receiving logs are kept on file for a minimum of 1 year.

Date Implemented: _____ By: _____

Date Reviewed: _____ By: _____

Date Revised: _____ By: _____

Receiving Log

Receiving Temperature Form

Date	Product	Temperature	Initials/Comments

Reviewed by: _____   Date: _____

RECEIVING LOG:

DATE	TIME	FOOD PRODUCT DESCRIPTION	PRODUCT CODE	CORRECTIVE ACTION TAKEN	EMPLOYEE INITIALS	MANAGER INITIALS

Receiving Reject

Receiving Reject Form

Date	Product	Rejected for:	Initials/Comments

Reviewed by: _____   **Date:** _____

Refrigeration/Cooler Log

Cooler Temperature Form

Date:	6:00 a.m. Temp.	10:00 a.m. Temp.	2:00 p.m. Temp.	6:00 p.m. Temp.	10:00 p.m. Temp.	Initials/Comments

Reviewed by: _____   Date: _____

COOKING LOG:

DATE	TIME	FOOD PRODUCT	INTERNAL TEMPERATURE °F/°C	CORRECTIVE ACTION TAKEN	EMPLOYEE INITIALS	MANAGER INITIALS

COOLING LOG PART 1:

DATE	FOOD	TIME	TEMP	+2 HOURS TIME	MUST BE 70°F (21.1°C) OR LOWER	+3 HOURS TIME	TEMP	+4 HOURS TIME	TEMP	+5 HOURS TIME	TEMP	+6 HOURS TIME	MUST BE 41°F (5°C) OR LOWER

COOLING—CORRECTIVE ACTION LOG PART 2:

DATE	FOOD PRODUCT	TIME	TEMPERATURE *MAX:* 70°F(21.1°C)— 2 hours *MAX:* 41°F(5°C)— 6 hours	CORRECTIVE ACTION TAKEN **Must:* Reheat **Must:* Discard	EMPLOYEE INITIALS	MANAGER INITIALS

Time-Temperature Log

Display Product Temperature Form

Product:	Product Temperatures					Initials
Time:						
7:00 a.m.						
9:00 a.m.						
11:00 a.m.						
1:00 p.m.						
3:00 p.m.						
5:00 p.m.						
7:00 p.m.						
9:00 p.m.						
11:00 p.m.						
1:00 a.m.						
3:00 a.m.						
5:00 a.m.						

Reviewed by: _____   **Date:** _____

Reheating

REHEATING LOG:

DATE	TIME	FOOD PRODUCT	INTERNAL TEMPERATURE	CORRECTIVE ACTION TAKEN	EMPLOYEE INITIALS	MANAGER INITIALS

HACCP Plan

FORM	HACCP Plan Form

Firm Name: _____   **Product Description:** _____

Firm Address: _____   **Method of Distribution and Storage:** _____

_____

_____

Intended Use and Consumer: _____

_____

_____

(1)	(2)	(3)	(4)	(5)	(6)	(7)	(8)	(9)	(10)
			Monitoring						
Critical Control Point (CCP)	Significant Hazard(s)	Critical Limits for each Preventive Measure	What	How	Frequency	Who	Corrective Action(s)	Records	Verification

Signature of Company Official: _____   **Date:** _____

Page 1 of 2

FORM (continued)

(1)	(2)	(3)	(4)	(5)	(6)	(7)	(8)	(9)	(10)
		Critical Limits for each Preventive Measure	Monitoring						
Critical Control Point (CCP)	Significant Hazard(s)		What	How	Frequency	Who	Corrective Action(s)	Records	Verification

Page 2 of 2

Source: Fish and Fisheries Products Hazards and Controls Guidance, Third Edition, June 2001, Center for Food Safety and Applied Nutrition, U.S. Food and Drug Administration, www.cfsan.fda.gov/~comm/haccp4x1.html.

Hazard Analysis Work Sheet

FORM	HACCP Analysis Work Sheet				

Firm Name:		Product Description:			
Firm Address:		Method of Distribution and Storage:			
		Intended Use and Consumer:			

(1)	(2)	(3)	(4)	(5)	(6)
Ingredient/Processing Step	Identify Potential Hazards Introduced, Controlled, or Enhanced at this Step (1)	Are any Potential Food Safety Hazards Significant? (Yes/No)	Justify your decision for Column 3	What Preventive Measure(s) can be Applied for the Significant Hazards?	Is this Step a Critical Control Point? (Yes/No)
	Biological				
	Chemical				
	Physical				
	Biological				
	Chemical				
	Physical				
	Biological				
	Chemical				
	Physical				
	Biological				
	Chemical				
	Physical				

Page 1 of 2

FORM (continued)

(1)	(2)	(3)	(4)	(5)	(6)
Ingredient/Processing Step	Identify Potential Hazards Introduced, Controlled, or Enhanced at this Step (1)	Are any Potential Food Safety Hazards Significant? (Yes/No)	Justify your decision for Column 3	What Preventive Measure(s) can be Applied for the Significant Hazards?	Is this Step a Critical Control Point? (Yes/No)
	Biological				
	Chemical				
	Physical				
	Biological				
	Chemical				
	Physical				
	Biological				
	Chemical				
	Physical				
	Biological				
	Chemical				
	Physical				
	Biological				
	Chemical				
	Physical				
	Biological				
	Chemical				
	Physical				

Page 2 of 2

Source: Fish and Fisheries Products Hazards and Controls Guidance, Third Edition, June 2001, Center for Food Safety and Applied Nutrition, U.S. Food and Drug Administration, www.cfsan.fda.gov/~comm/haccp4x1.html.

Process 1—Food Preparation with No-Cook Step

Process #1—Food Preparation with No-Cook Step

Food/Menu Items:

Hazard(s)	Critical Control Points (List Only the Operational Steps That Are CCPs)	Critical Limits	Monitoring	Corrective Actions	Verification	Records
Prerequisite Programs						

Process #1—Food Preparation with No-Cook Step

Food/Menu Items:

Process Step	Hazard(s)	CCP (Y/N)	Critical Limits	Monitoring	Corrective Actions	Verification	Records
RECEIVE							
STORE							
PREPARE							
HOLD							
SERVE							
Prerequisite Programs							

Source: Managing Food Safety: A Regulator's Manual for Applying HACCP Principles to Risk-based Retail and Food Service Inspections and Evaluating Voluntary Food Safety Management Systems, Center for Food Safety and Applied Nutrition, U.S. Food and Drug Administration, www.cfsan.fda.gov/~dms/hret3-a2.html.

Process 2—Preparation for Same-Day Service

Process #2—Preparation for Same-Day Service

Food/Menu Items:

Hazard(s)	Critical Control Points (List Only the Operational Steps That Are CCPs)	Critical Limits	Monitoring	Corrective Actions	Verification	Records
Prerequisite Programs						

Process #2—Preparation for Same-Day Service

Food/Menu Items:

Process Step	Hazard(s)	CCP (Y/N)	Critical Limits	Monitoring	Corrective Actions	Verification	Records
RECEIVE							
STORE							
PREPARE							
COOK							
HOLD							
SERVE							
Prerequisite Programs							

Source: Managing Food Safety: A Regulator's Manual for Applying HACCP Principles to Risk-based Retail and Food Service Inspections and Evaluating Voluntary Food Safety Management Systems, Center for Food Safety and Applied Nutrition, U.S. Food and Drug Administration, www.cfsan.fda.gov/~dms/hret3-a2.html.

Process 3—Complex Food Preparation

Process #3—Complex Food Preparation

Food/Menu Items:

Hazard(s)	Critical Control Points (List Only the Operational Steps That Are CCPs)	Critical Limits	Monitoring	Corrective Actions	Verification	Records
Prerequisite Programs						

Process #3—Complex Food Preparation

Food/Menu Items:

Process Step	Hazard(s)	CCP (Y/N)	Critical Limits	Monitoring	Corrective Actions	Verification	Records
RECEIVE							
STORE							
PREPARE							
COOK							
COOL							
REHEAT							
HOLD							
SERVE							
Prerequisite Programs							

Source: Managing Food Safety: A Regulator's Manual for Applying HACCP Principles to Risk-based Retail and Food Service Inspections and Evaluating Voluntary Food Safety Management Systems, Center for Food Safety and Applied Nutrition, U.S. Food and Drug Administration, www.cfsan.fda.gov/~dms/hret3-a2.html.

Thermometer Calibration

THERMOMETER CALIBRATION LOG:

Date	Manager	Employee	AM Time	MID Time	PM Time	Date	Manager	Employee	AM Time	MID Time	PM Time

Waste/Shrink/Discard Chart

DISCARD LOG:

DATE	CODE	FOOD PRODUCT DESCRIPTION	• HOLD • DISCARD • RETURN • CREDIT	EXPLAIN ACTION TAKEN REASON	EMPLOYEE INITIALS	MANAGER INITIALS

HACCP Monitoring Report

FORM	Hazard Analysis Critical Control Point Monitoring Procedure Report

New York State Department of Health
Bureau of Community Sanitation and Food Protection

COUNTY	DIST.				EST. NO.				MONTH		DAY		YEAR	

THIS FORM CONSISTS OF TWO PAGES AND BOTH MUST BE COMPLETED.

Establishment Name: **Operator's Name:**

Address:

(T) (C) (V)   **County:**

Food

Process (Step) Circle CCPs	Criteria for Control	Monitoring Procedure or What to Look For	Actions to Take When Criteria Not Met
Receiving/Storage	☐ Approved source (inspected) ☐ Shellfish tags ☐ Raw/Cooked/Separated in storage ☐ Refrigerate at less than or equal to 40°F (4.4°C)	☐ Shellfish tags available ☐ Shellfish tags complete ☐ Measure food temperature ☐ No raw foods stored above cooked or ready-to-eat foods	☐ Discard food ☐ Return food ☐ Separate raw from cooked food ☐ Discard cooked food contaminated by raw food ☐ Food Temperature more than 40°F (4.4°C) more than 2 hours, discard food ☐ More than 70°F (21.1°C), discard food
Thawing	☐ Under refrigeration ☐ Under running water less than 70°F (21.1°C) ☐ Microwave ☐ Less than 3 lbs., cooked frozen ☐ More than 3 lbs., do not cook until thawed	Observe method Measure food temperature	Food temperature More than or equal to 70°F (21.1°C), discard More than 40°F (4.4°C) more than 2 hours, discard
Processing Prior to Cooking	Food temperature less than or equal to 45°F (7.2°C)	Observe quantity of food at room temperature Observe time food held at room temperature	Food temperature: More than 40°F (4.4°C) more than 2 hours, discard food More than 70°F (21.1°C), discard food
Cooking	Temperature to kill pathogens Food temperature at thickest part more than or equal to ____°F ____°C	Measure food temperature at thickest part	Continue cooking until food temperature at thickest part is more than equal to ____°F ____°C

FORM (continued)

Process (Step) Circle CCPs	Criteria for Control	Monitoring Procedure or What to Look For	Actions to Take When Criteria Not Met
Hot Holding	Food temperature at thickest part more than or equal to ____°F ____°C	Measure food temperature at thickest part during hot holding every ____ minutes	Food temperature: 140°F–120°F (60°C–48.9°C) More than or equal to 2 hours, discard; less than 2 hours, reheat to 165°F (73.9°C) and hold at 140°F (60°C) 120°F–40°F More than or equal to 2 hours, discard; less than 2 hours, reheat to 165°F (73.9°C) and hold at 140°F (60°C)
Cooling	Food 120°F to 70°F (48.9°C to 21.2°C) in hours: 70°F to 40°F (21.1°C to 4.4°C) in 4 additional hours by the following methods: (check all that apply) ☐ Product depth less than or equal to 3 inches ☐ Ice water bath and stirring ☐ Solid pieces less than or equal to 6 lbs. ☐ Rapid chill all ingredients ☐ No covers until cold	Measure temperature during cooling every ____ minutes ☐ Food depth ☐ Food iced ☐ Food stirred ☐ Food size ☐ Food placed in rapid chill refrigeration unit ☐ Food uncovered	Food temperature: 120°F–70°F (48.9°C to 21.1°C) more than 2 hours, discard food 70°F–40°F (21.1°C to 4.4°C) more than 4 hours, discard food 40°F (4.4°C) or less but cooled too slowly, discard food
Processing/Slicing/Deboning/ Mixing/Dicing/Assembling/ Serving	Prevent contamination by: Ill workers not working Workers hands not touching ready-to-eat foods Workers hands washed Cold potentially hazardous food at temperatures less than or equal to 140°F (60°C) Hot potentially hazardous food at temperature more than or equal to 140°F (60°C) Equipment and utensils clean and sanitized	Observe: Workers' health Use of gloves, utensils Handwashing technique Wash & sanitize equipment & utensils Use prechilled ingredients for cold foods Minimize quantity of food at room temperature Measure food temperature	If yes to following, discard: Ill worker working Direct hand contact with ready-to-eat food observed Cold potentially hazardous food: more than 40°F (4.4°C) more than or equal to 2 hours, discard; more than 70°F (21.1°C), discard Hot potentially hazardous food 140°F–120°F (60°C–48.9°C) More than or equal to 2 hours, discard; less than 2 hours, reheat to 165°F (73.9°C) and hold at 140°F (60°C) If yes to following, discard or reheat to 165°F (73.9°C) Raw food contaminated by other food Equipment/utensils are contaminated

FORM (continued)

Process (Step) Circle CCPs	Criteria for Control	Monitoring Procedure or What to Look For	Actions to Take When Criteria Not Met
Reheating	Food temperature at thickest part more than or equal to 165°F (73.9°C)	Measure food temperature during reheating	Food temperature less than 165°F (73.9°C), continue reheating
Holding Food, Hot/Cold Transporting Food	Food temperature ☐ More than or equal to 140°F (60°C) at thickest part ☐ Less than or equal to 40°F (4.4°C) at thickest part	Measure food temperature during holding every ____ minutes	☐ Hot holding potentially hazardous food: 140°F–120°F (60°C–48.9°C) More than or equal to 2 hours, discard; less than 2 hours, reheat to 165°F (73.9°C) and hold at 140°F (60°C) 120°F–40°F (48.9°C–4.4°C) More than or equal to 2 hours, discard; less than 2 hours, reheat to 165°F (73.9°C) and hold at 140°F (60°C) ☐ Cold holding potentially hazardous food temperature: 40°F–70°F (4.4°C–21.1°C) More than or equal to 2 hours, discard; less than 2 hours, serve or refrigerate More than equal to 70°F (21.2°C), discard

I have read the above food preparation procedures and agree to follow and monitor the critical control points and to take appropriate corrective action when needed. If I want to make any changes, I will notify the Health Department prior to such a change.

Signature of person in charge _____   Signature of inspector _____

FBI Investigation

FORM	Preliminary Foodborne Illness Investigation*

Restaurant Name:
Address:
Suspected Food/Beverages (be specific):

Date Meal Eaten:	**Time:**	**Onset Date:**	**Onset Time:**
Caller's Name:		**Phone:**	**Incub Time:**
Address:			

Person's Name Who Became Ill:	**Name/Age/Sex/Occupation**

Dcotor Seen: ☐ Yes ☐ No	**Diagnosis/Lab Results:**
Clinic Name/Doctor's Name:	
Address:	**Phone:**
☐ Stool ☐ Blood Other:	**Results:**
Date Received:	**Call Received by:**

SYMPTOMS

☐ Vomiting	☐ Fever	☐ Nausea	☐ Burning Mouth
__ No. of Days	☐ Chills	☐ Muscle Ache	☐ Itching
__ No. of Times	☐ Cramps	☐ Excess Salivation	☐ Rash
☐ Diarrhea	☐ Headache	☐ Cough	☐ Dizziness
__ No. Days	☐ Perspiration	☐ Metallic Taste	☐ Numbness
__ No. of Times			☐ Double Vision
☐ Bloody	Other _____		
☐ Explosive	_____		
☐ Watery			

FORM (continued)

FOOD HISTORY

1st 24 Hours/Date Meal Consumed	All Foods Consumed	Where
Dinner		
Lunch		
Breakfast		

2nd 24 Hours (Previous Day)

Dinner		
Lunch		
Breakfast		

3rd 24 Hours (2 Days Prior)   All Foods Consumed at Restaurants or From Caterers

Dinner	
Lunch	
Breakfast	

1. Was this a take-out order?
2. (If Yes) Elapsed time between pickup to consumption _____ HRS
3. Are there any other ill contacts (including pets)? _____
4. If yes to #4, please list symptoms _____
5. Please note anything unusual noticed about meal? (temperature, taste, etc.) _____

(PLEASE USE REVERSE SIDE IF MORE ROOM IS NEEDED)

*Denver Public Health Department, Consumer Protection.

Food Safety Checklist

FORM	**Food Safety Checklist**

Date _____   Observer _____

Directions: Use this checklist daily to determine areas in your operations requiring corrective action. Record corrective action taken and keep completed records in a notebook for future reference.

Personal Hygiene

	Yes	No	Corrective Action
• Employees wear clean and proper uniform including shoes.	☐	☐	_____
• Effective hair restraints are properly worn.	☐	☐	_____
• Fingernails are short, unpolished, and clean (no artificial nails).	☐	☐	_____
• Jewelry is limited to a plain ring, such as a wedding band and a watch—no bracelets.	☐	☐	_____
• Hands are washed properly, frequently, and at appropriate times.	☐	☐	_____
• Burns, wounds, sores or scabs, or splints and water-proof bandages on hands are bandaged and completely covered with a foodservice glove while handling food.	☐	☐	_____
• Eating, drinking, chewing gum, smoking, or using tobacco are allowed only in designated areas away from preparation, service, storage, and ware washing areas.	☐	☐	_____
• Employees use disposable tissues when coughing or sneezing and then immediately wash hands.	☐	☐	_____
• Employees appear in good health.	☐	☐	_____
• Hand sinks are unobstructed, operational, and clean.	☐	☐	_____
• Hand sinks are stocked with soap, disposable towels, and warm water.	☐	☐	_____
• A handwashing reminder sign is posted.	☐	☐	_____
• Employee restrooms are operational and clean.	☐	☐	_____

Food Preparation

	Yes	No	Corrective Action
• All food stored or prepared in facility is from approved sources.	☐	☐	_____
• Food equipment utensils, and food contact surfaces are properly washed, rinsed, and sanitized before every use.	☐	☐	_____
• Frozen food is thawed under refrigeration or in cold running water.	☐	☐	_____
• Preparation is planned so ingredients are kept out of the temperature danger zone to the extent possible.	☐	☐	_____
• Food is tasted during the proper procedure.	☐	☐	_____
• Procedures are in place to prevent cross-contamination.	☐	☐	_____
• Food is handled with suitable utensils, such as single use gloves or tongs.	☐	☐	_____
• Food is prepared in small batches to limit the time it is in the temperature danger zone.	☐	☐	_____
• Clean reusable towels are used only for sanitizing equipment, surfaces and not for drying hands, utensils, or floor.	☐	☐	_____
• Food is cooked to the required safe internal temperature for the appropriate time. The temperature is tested with a calibrated food thermometer.	☐	☐	_____
• The internal temperature of food being cooked is monitored and documented.	☐	☐	_____

Hot Holding

	Yes	No	Corrective Action
• Hot holding unit is clean.	☐	☐	_____
• Food is heated to the required safe internal temperature before placing in hot holding. Hot holding units are not used to reheat potentially hazardous food.	☐	☐	_____
• Hot holding unit is pre-heated before hot food is placed in unit.	☐	☐	_____
• Temperature of hot food being held is at or above 135°F.	☐	☐	_____
• Food is protected from contamination.	☐	☐	_____

FORM (continued)

Cold Holding

	Yes	No	Corrective Action
• Refrigerators are kept clean and organized.	☐	☐	_____
• Temperature of cold food being held is at or below 41°F.	☐	☐	_____
• Food is protected from contamination.	☐	☐	_____

Refrigerator, Freezer, and Milk Cooler

	Yes	No	Corrective Action
• Thermometers are available and accurate.	☐	☐	_____
• Temperature is appropriate for pieces of equipment.	☐	☐	_____
• Food is stored 6 inches off floor or in walk-in cooling equipment.	☐	☐	_____
• Refrigerator and freezer units are clean and neat.	☐	☐	_____
• Proper chilling procedures are used.	☐	☐	_____
• All food is properly wrapped, labeled, and dated.	☐	☐	_____
• The FIFO (First In, First Out) method of inventory management is used.	☐	☐	_____
• Ambient air temperature of all refrigerators and freezers is monitored and documented at the beginning and end of each shift.	☐	☐	_____

Food Storage and Dry Storage

	Yes	No	Corrective Action
• Temperatures of dry storage area is between 50°F and 70°F (10°C and 21.1°C) or State public health department requirement.	☐	☐	_____
• All food and paper supplies are stored 6 to 8 inches off the floor.	☐	☐	_____
• All food is labeled with name and received date.	☐	☐	_____
• Open bags of food are stored in containers with tight fitting lids and labeled with common name.	☐	☐	_____
• The FIFO (First In, First Out) method of inventory management is used.	☐	☐	_____
• There are no bulging or leaking canned goods.	☐	☐	_____
• Food is protected from contamination.	☐	☐	_____
• All food surfaces are clean.	☐	☐	_____
• Chemicals are clearly labeled and stored away from food and food related supplies.	☐	☐	_____
• There is a regular cleaning schedule for all food surfaces.	☐	☐	_____

Cleaning and Sanitizing

	Yes	No	Corrective Action
• Three-compartment sink is properly set up for ware washing.	☐	☐	_____
• Dishmachine is working properly (i.e. gauges and chemicals are at recommended levels).	☐	☐	_____
• Water is clean and free of grease and food particles.	☐	☐	_____
• Water temperatures are correct for wash and rinse.	☐	☐	_____
• If heat sanitizing the utensils are allowed to remain immersed in 171°F (77.2°C) water for 30 seconds.	☐	☐	_____
• If using a chemical sanitizer, it is mixed correctly and a sanitizer strip is used to test chemical concentration.	☐	☐	_____
• Smallware and utensils are allowed to air dry.	☐	☐	_____
• Wiping cloths are stored in sanitizing solution while in use.	☐	☐	_____

Utensils and Equipment

	Yes	No	Corrective Action
• All small equipment and utensils, including cutting boards and knives, are cleaned and sanitized between uses.	☐	☐	_____
• Small equipment and utensils are washed, sanitized, and air-dried.	☐	☐	_____
• Work surfaces and utensils are clean.	☐	☐	_____
• Work surfaces are cleaned and sanitized between uses.	☐	☐	_____

FORM (continued)

Utensils and Equipment (cont.)

	Yes	No	Corrective Action
• Thermometers are cleaned and sanitized after each use.	☐	☐	_____
• Thermometers are calibrated on a routine basis.	☐	☐	_____
• Can opener is clean.	☐	☐	_____
• Drawers and racks are clean.	☐	☐	_____
• Clean utensils are handled in a manner to prevent contamination of areas that will be in direct contact with food or a person's mouth.	☐	☐	_____

Large Equipment

	Yes	No	Corrective Action
• Food slicer is clean.	☐	☐	_____
• Food slicer is broken down, cleaned, and sanitized before and after every use.	☐	☐	_____
• Boxes, containers, and recyclables are removed from site.	☐	☐	_____
• Loading dock and area around dumpsters are clean and odor-free.	☐	☐	_____
• Exhaust hood and filters are clean.	☐	☐	_____

Garbage Storage and Disposal

	Yes	No	Corrective Action
• Kitchen garbage cans are clean and kept covered.	☐	☐	_____
• Garbage cans are emptied as necessary.	☐	☐	_____
• Boxes and containers are removed from site.	☐	☐	_____
• Loading dock and area around dumpster are clean.	☐	☐	_____
• Dumpsters are clean.	☐	☐	_____

Pest Control

	Yes	No	Corrective Action
• Outside doors have screens, are well-sealed, and are equipped with a self-closing device.	☐	☐	_____
• No evidence of pests is present.	☐	☐	_____
• There is a regular schedule of pest control by licensed pest control operator.	☐	☐	_____

272 HACCP Star Point 5

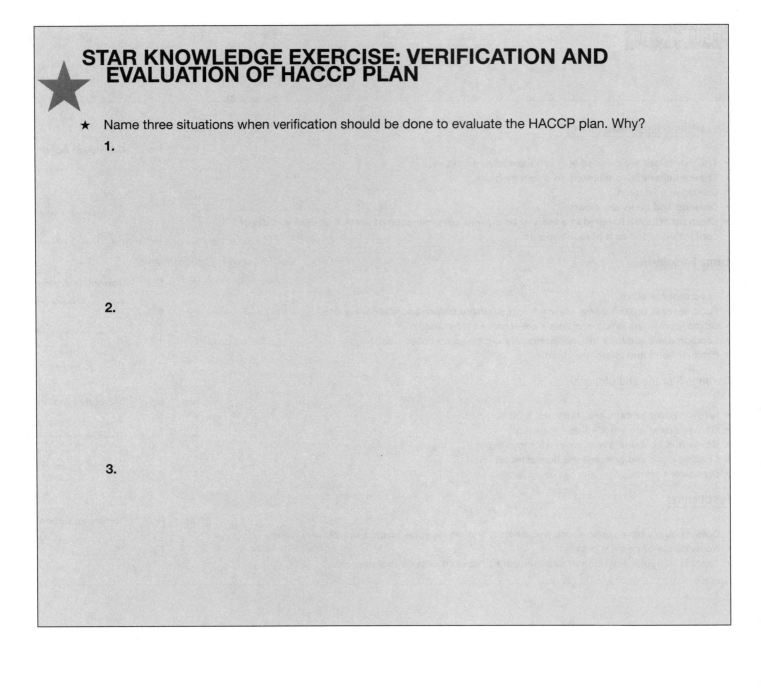

STAR KNOWLEDGE EXERCISE: VERIFICATION AND EVALUATION OF HACCP PLAN

★ Name three situations when verification should be done to evaluate the HACCP plan. Why?

1.

2.

3.

HACCP PRINCIPLES MATCH GAME

Here is HACCP at a glance. Match the principle with the picture.

_____   _____   _____

A. Identify Corrective Actions

B. Determine Critical Control Points

C. Verify That the System Works

D. Establish Monitoring Procedures

E. Establish Procedures for Record Keeping and Documentation

F. Conduct a Hazard Analysis

G. Establish Critical Limits

_____   _____   _____

_____

HACCP Principles Match Game

How many points did you earn? _____

★ **If you scored 7 points—Congratulations! You are a HACCP Principles Superstar!**

★ **If you scored 5–6 points**—Good job! You have a basic understanding of HACCP principles.

★ **If you scored 3–4 points**—It is critical to go back and review the 7 HACCP principles featured in Star Points 3, 4, and 5.

★ **If you scored 0–2 points**—The time to review is now! Please do not proceed. Go back and review the 7 HACCP principles featured in Star Points 3, 4, and 5. Your trainer will help you in this process.

In Star Point 5, we have discussed the checks and balances system of your HACCP plan—verification and record keeping. Verification **confirms** your plan is working properly. Record keeping is **proof** that your HACCP plan is working properly and that you're achieving active managerial control of foodborne risk factors. This is the final step in becoming a HACCP Superstar. Now that your HACCP Star is complete, you should experience the benefits of all these points. First is the health and safety of your customers and employees. Second is an overall improvement of your foodservice operation, including food safety SOPs, food quality, cleanliness, sanitation, food defense SOPs, and a team better equipped to handle food. Finally, the HACCP Star took you through the process of analyzing and controlling hazards. You now manage a foodservice operation that is proactive versus reactive. You will do your part with confidence to ensure the overall safety of our food supply from farm to table.

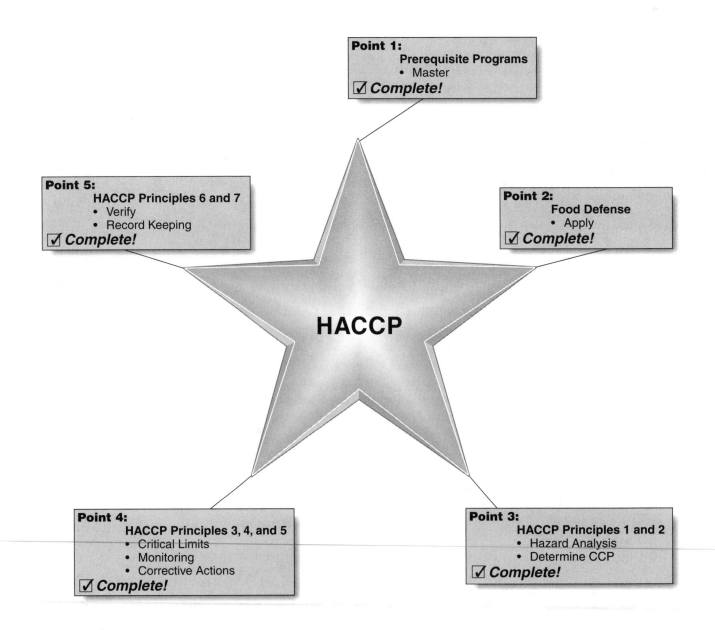

⭐ ARE YOU A HACCP "SUPERSTAR"?

The goal of this manual is to help you to better understand the five points in the HACCP Star and to demonstrate how they **save lives!** Each point of the HACCP Star helps you to create and use an effective HACCP plan. Now that you have a better understanding of a HACCP plan, it is up to you to be a valued **HACCP team member** by applying the standard operating procedures and HACCP principles learned in this book to your foodservice operation and leading your fellow team members to make these steps a part of your daily routines. Your leadership will help to ensure that you're serving and selling safe food!

Time/Temperature Control for Safety Food (TCS)

In this section, you will find the four tables needed for time/temperature control for safety of food (TCS). Table 1 gives approximate water activity for pathogens. Table 2 provides approximate water activity for selected food categories. Table 3 gives approximate pH values permitting the growth of selected pathogens in food. Table 4 provides approximate pH values of common foods. Additional information can be found at www.cfsan.fda.gov/~comm/lacf-phs.

Table 1   Approximate a_w Values for Growth of Selected Pathogens in Food

Organism	Minimum	Optimum	Maximum
Campylobacter spp.	0.98	0.99	
Clostridium botulinum type E	0.97		
Clostridium botulinum types A and B	0.93		
Shigella spp.	0.97		
Yersinia enterocolitica	0.97		
Enterohemorrhagic *Escherichia coli*	0.95	0.99	
Vibrio vulnificus	0.96	0.98	0.99
Salmonella spp.	0.94	0.99	>0.99
Vibrio parahaemolyticus	0.94	0.98	0.99
Bacillus cereus	0.93		
Clostridium perfringens	0.943	0.95–0.96	0.97
Listeria monocytogenes	0.92		
Staphylococcus aureus growth	0.83	0.98	0.99
Staphylococcus aureus toxin	0.88	0.98	0.99

Table 2   Approximate Water Activity of Selected Food Categories

Animal Products	a_w
Fresh meat, poultry, fish	0.99–1.00
Natural cheeses	0.95–1.00
Pudding	0.97–0.99
Eggs	0.97
Cured meat	0.87–0.95
Sweetened condensed milk	0.83
Parmesan cheese	0.68–0.76
Honey	0.75
Dried whole egg	0.40
Dried whole milk	0.20
Plant Products	a_w
Fresh fruits, vegetables	0.97–1.00
Bread	~0.96
Bread, white	0.94–0.97
Bread, crust	0.30
Baked cake	0.90–0.94
Maple syrup	0.85
Jam	0.75–0.80
Jellies	0.82–0.94
Uncooked rice	0.80–0.87
Fruit juice concentrates	0.79–0.84
Fruit cake	0.73–0.83
Cake icing	0.76–0.84
Flour	0.67–0.87
Dried fruit	0.55–0.80
Cereal	0.10–0.20
Sugar	0.19
Crackers	0.10

Table 3 Approximate pH Values Permitting the Growth of Selected Pathogens in Food

Microorganism	Minimum	Optimum	Maximum
Campylobacter spp.	4.9	6.5–7.5	9.0
Clostridium botulinum toxin	4.6		8.5
Clostridium botulinum growth	4.6		8.5
Shigella spp.	4.9		9.3
Yersinia enterocolitica	4.2	7.2	9.6
Enterohemorrhagic *Escherichia coli*	4.4	6.0–7.0	9.0
Vibrio vulnificus	5.0	7.8	10.2
Salmonella spp.	4.2[1]	7.0–7.5	9.5
Vibrio parahaemolyticus	4.8	7.8–8.6	11.0
Bacillus cereus	4.9	6.0–7.0	8.8
Clostridium perfringens	5.5–5.8	7.2	8.0–9.0
Listeria monocytogenes	4.39	7.0	9.4
Staphylococcus aureus growth	4.0	6.0–7.0	10.0
Staphylococcus aureus toxin	4.5	7.0–8.0	9.6

Table 4   Approximate pH Values of Selected Food Categories

Item	Approximate pH
Abalone	6.10–6.50
Abalone mushroom	5.00
Anchovies	6.50
Anchovies, stuffed w/capers, in olive oil	5.58
Antipasto	5.60
Apple, baked with sugar	3.20–3.55
Apple, eating	3.30–4.00
Apples	
Delicious	3.90
Golden Delicious	3.60
Jonathan	3.33
McIntosh	3.34
Juice	3.35–4.00
Sauce	3.10–3.60
Winesap	3.47
Apricots	3.30–4.80
Canned	3.40–3.78
Dried, stewed	3.30–3.51
Nectar	3.78
Pureed	3.42–3.83
Strained	3.72–3.95
Arrowroot crackers	6.63–6.80
Arrowroot cruel	6.37–6.87
Artichokes	5.50–6.00
Artichokes, canned, acidified	4.30–4.60
Artichokes, French, cooked	5.60–6.00
Artichokes, Jerusalem, cooked	5.93–6.00
Asparagus	6.00–6.70
Buds	6.70
Stalks	6.10
Asparagus, cooked	6.03–6.16
Asparagus, canned	5.00–6.00
Asparagus, frozen, cooked	6.35–6.48
Asparagus, green, canned	5.20–5.32
Asparagus, strained	4.80–5.09
Avocados	6.27–6.58
Baby corn	5.20
Baby food soup, unstrained	5.95–6.05
Bamboo shoots +	5.10–6.20
Bamboo Shoots, preserved	3.50–4.60

(continues)

Table 4   Continued

Item	Approximate pH
Bananas	4.50–5.20
Bananas, red	4.58–4.75
Bananas, yellow	5.00–5.29
Barley, cooked	5.19–5.32
Basil pesto	4.90
Bass, sea, broiled	6.58–6.78
Bass, striped, broiled	6.50–6.70
Beans	5.60–6.50
Black	5.78–6.02
Boston style	5.05–5.42
Kidney	5.40–6.00
Lima	6.50
Soy	6.00–6.60
String	5.60
Wax	5.30–5.70
Beans, pork and tomato sauce, canned	5.10–5.80
Beans, refried	5.90
Beans, vegetarian, tomato sauce, canned	5.32
Beets	5.30–6.60
Beets, cooked	5.23–6.50
Beets, canned, acidified	4.30–4.60
Beets, canned	4.90–5.80
Beets, chopped	5.32–5.56
Beets, strained	5.32–5.56
Blackberries, Washington	3.85–4.50
Blueberries, Maine	3.12–3.33
Blueberries, frozen	3.11–3.22
Bluefish, Boston, filet, broiled	6.09–6.50
Bran	
Flakes	5.45–5.67
All-Bran	5.59–6.19
Bread, white	5.00–6.20
Bread, Boston, brown	6.53
Bread, cracked wheat	5.43–5.50
Bread, pumpernickel	5.40
Bread, rye	5.20–5.90
Bread, whole wheat	5.47–5.85
Breadfruit, cooked	5.33

Table 4 Continued

Item	Approximate pH
Broccoli, cooked	6.30–6.52
Broccoli, frozen, cooked	6.30–6.85
Broccoli, canned	5.20–6.00
Brussels sprout	6.00–6.30
Buttermilk	4.41–4.83

Cabbage
  Green
  Red
  Savoy
  White
	5.20–6.80
	5.50–6.75
	5.60–6.00
	6.30
	6.20

Cactus	4.70
Calamari (squid)	5.80
Cantaloupe	6.13–6.58
Capers	6.00
Carp	6.00

Carrots
  Carrots, canned
  Carrots, chopped
  Carrots, cooked
  Carrots, pureed
  Carrots, strained
	5.88–6.40
	5.18–5.22
	5.30–5.56
	5.58–6.03
	4.55–5.80
	5.10

Cauliflower	5.60
Cauliflower, cooked	6.45–6.80
Caviar, American	5.70–6.00

Celery
  Celery, cooked
  Celery knob, cooked
	5.70–6.00
	5.37–5.92
	5.71–5.85

Cereal, strained	6.44–6.45

Cheese, American, mild
  Cheese, Camembert
  Cheese, cheddar
  Cheese, cottage
  Cheese, cream
  Cheese dip
  Cheese, Edam
  Cheese, Old English
  Cheese, Roquefort
  Cheese, parmesan
  Cheese, snippy
	4.98
	7.44
	5.90
	4.75–5.02
	4.10–4.79
	5.80
	5.40
	6.15
	5.10–5.98
	5.20–5.30
	5.18–5.21

(continues)

Table 4   Continued

Item	Approximate pH
Cheese, Stilton	5.70
Cheese, Swiss Gruyere	5.68–6.62
Cherries, California	4.01–4.54
Cherries, frozen	3.32–3.37
Cherries, black, canned	3.82–3.93
Cherries, maraschino	3.47–3.52
Cherries, red, water-pack	3.25–3.82
Cherries, Royal Ann	3.80–3.83
Chicory	5.90–6.05
Chili sauce, acidified	2.77–3.70
Chives	5.20–6.31
Clams	6.00–7.10
Clam chowder, New England	6.40
Coconut, fresh	5.50–7.80
Coconut milk	6.10–7.00
Coconut preserves	3.80–7.00
Codfish, boiled	5.30–6.10
Cod liver	6.20
Conch	7.52–8.40
Corn	5.90–7.30
Corn, canned	5.90–6.50
Cornflakes	4.90–5.38
Corn, frozen, cooked	7.33–7.68
Corn, Golden Bantam, cooked on cob	6.22–7.04
Crabmeat	6.50–7.00
Crabapple jelly, corn	2.93–3.02
Cranberry juice, canned	2.30–2.52
Crabmeat, cooked	6.62–6.98
Cream, 20 per cent	6.50–6.68
Cream, 40 per cent	6.44–6.80
Cream of asparagus	6.10
Cream of coconut, canned	5.51–5.87
Cream of potato soup	6.00
Cream of Wheat, cooked	6.06–6.16
Chrysanthemum drink	6.50
Cucumbers	5.12–5.78
Cucumbers, dill pickles	3.20–3.70
Cucumbers, pickled	4.20–4.60

Table 4 Continued

Item	Approximate pH
Curry sauce	6.00
Curry paste, acidified	4.60–4.80
Cuttlefish	6.30
Dates, canned	6.20–6.40
Dates, dromedary	4.14–4.88
Dungeness crabmeat	
Eggplant	5.50–6.50

Eggs, new-laid, whole
    White
    Yolk
	6.58
	7.96
	6.10

Eel	6.20
Escarolle	5.70–6.00
Enchilada sauce	4.40–4.70
Fennel (anise)	5.48–5.88
Fennel, cooked	5.80–6.02
Figs, calamyrna	5.05–5.98
Figs, canned	4.92–5.00
Flounder, boiled	6.10–6.90
Flounder, filet, broiled	6.39–6.89
Four bean salad	5.60
Fruit cocktail	3.60–4.00
Garlic	5.80
Gelatin dessert	2.60
Gelatin, plain	6.08
Gherkin	
Ginger	5.60–5.90
Ginseng, Korean drink	6.00–6.50
Gooseberries	2.80–3.10
Graham crackers	7.10–7.92

Grapes, canned
    Grapes, Concord
    Grapes, Lady Finger
	3.50–4.50
	2.80–3.00
	3.51–3.58

(continues)

Table 4   Continued

Item	Approximate pH
Grapes, Malaga	3.71–3.78
Grapes, Niagara	2.80–3.27
Grapes, Ribier	3.70–3.80
Grapes, seedless	2.90–3.82
Grapes, Tokyo	3.50–3.84
Grapefruit	3.00–3.75
Grapefruit, canned	3.08–3.32
Grapefruit juice, canned	2.90–3.25
Grass jelly	5.80–7.20
Greens, mixed, chopped	5.05–5.22
Greens, mixed, strained	5.22–5.30
Grenadine syrup	2.31
Guava nectar	5.50
Guava, canned	3.37–4.10
Guava jelly	3.73
Haddock, filet, broiled	6.17–6.82
Hearts of palm	5.70
Herring	6.10
Hominy, cooked	6.00–7.50
Honey	3.70–4.20
Honey aloe	4.70
Horseradish, freshly ground	5.35
Huckleberries, cooked with sugar	3.38–3.43
Jackfruit	4.80–6.80
Jam, fruit	3.50–4.50
Jellies, fruit	3.00–3.50
Jujube	5.20
Junket-type dessert:	
Raspberry	6.27
Vanilla	6.49
Kale, cooked	6.36–6.80
Ketchup	3.89–3.92
Kippered, herring, Marshall	5.75–6.20
Herring, pickled	4.50–5.00

Table 4 Continued

Item	Approximate pH
Kelp	6.30
Kumquat, Florida	3.64–4.25
Leeks Leeks, cooked	5.50–6.17 5.49–6.10
Lemon juice	2.00–2.60
Lentils, cooked	6.30–6.83
Lentil soup	5.80
Lettuce	5.80–6.15
Lettuce, Boston	5.89–6.05
Lettuce, iceberg	5.70–6.13
Lime juice	2.00–2.35
Lime	2.00–2.80
Lobster bisque	6.90
Lobster soup	5.70
Lobster, cooked	7.10–7.43
Loganberries	2.70–3.50
Loquat (may be acidified to pH 3.8)	5.10
Lotus root	6.90
Lychee	4.70–5.01
Macaroni, cooked	5.10–6.41
Mackerel, king, broiled	6.26–6.50
Mackerel, Spanish, broiled	6.07–6.36
Mackerel, canned	5.90–6.40
Mangoes, ripe	3.40–4.80
Mangoes, green	5.80–6.00
Maple syrup	5.15
Maple syrup, light (acidified)	4.60
Matzos	5.70
Mayhaw (a variety of strawberry)	3.27–3.86
Melba toast	5.08–5.30
Melon, casaba	5.78–6.00

(continues)

Table 4   Continued

Item	Approximate pH
Melons, honeydew	6.00–6.67
Melons, Persian	5.90–6.38
Milk, cow	6.40–6.80
Milk, acidophilus	4.09–4.25
Milk, condensed	6.33
Milk, evaporated	5.90–6.30
Milk, goat's	6.48
Milk, peptonized	7.10
Milk, sour, fine curd	4.70–5.65
Milkfish	5.30
Mint jelly	3.01
Molasses	4.90–5.40
Muscadine (a variety of grape)	3.20–3.40
Mushrooms	6.00–6.70
Mushrooms, cooked	6.00–6.22
Mushroom soup, cream of, canned	5.95–6.40
Mussels	6.00–6.85
Mustard	3.55–6.00
Nata de coco	5.00
Nectarines	3.92–4.18
Noodles, boiled	6.08–6.50
Oatmeal, cooked	6.20–6.60
Octopus	6.00–6.50
Okra, cooked	5.50–6.60
Olives, black	6.00–7.00
Olives, green, fermented	3.60–4.60
Olives, ripe	6.00–7.50
Onions, pickled	3.70–4.60
Onions, red	5.30–5.80
Onions, white	5.37–5.85
Onions, yellow	5.32–5.60
Oranges, Florida	3.69–4.34
Oranges, Florida "color added"	3.60–3.90
Orange juice, California	3.30–4.19
Orange juice, Florida	3.30–4.15
Orange marmalade	3.00–3.33

Table 4   Continued

Item	Approximate pH
Oysters	5.68–6.17
Oyster, smoked	6.00
Oyster mushrooms	5.00–6.00
Palm, heart of	6.70
Papaya	5.20–6.00
Papaya marmalade	3.53–4.00
Parsley	5.70–6.00
Parsnip	5.30–5.70
Parsnips, cooked	5.45–5.65
Pâté	5.90
Peaches	3.30–4.05
Peaches, canned	3.70–4.20
Peaches, cooked with sugar	3.55–3.72
Peaches, frozen	3.28–3.35
Peanut butter	6.28
Peanut soup	7.5
Pears, Bartlett	3.50–4.60
Pears, canned	4.00–4.07
Pears, sickle cooked w/sugar	4.04–4.21
Pear nectar	4.03
Peas, canned	5.70–6.00
Peas, chick, garbanzo	6.48–6.80
Peas, cooked	6.22–6.88
Peas, dried (split green), cooked	6.45–6.80
Peas, dried (split yellow), cooked	6.43–6.62
Peas, frozen, cooked	6.40–6.70
Peas, pureed	4.90–5.85
Pea soup, cream of, canned	5.70
Peas, strained	5.91–6.12
Peppers	4.65–5.45
Peppers, green	5.20–5.93
Persimmons	4.42–4.70
Pickles, fresh pack	5.10–5.40
Pimento	4.40–4.90
Pimento, canned, acidified	4.40–4.60
Pineapple	3.20–4.00
Pineapple, canned	3.35–4.10
Pineapple juice, canned	3.30–3.60

(continues)

Table 4   Continued

Item	Approximate pH
Plum Nectar	3.45
Plums, blue	2.80–3.40
Plums, Damson	2.90–3.10
Plums, frozen	3.22–3.42
Plums, Green Gage	3.60–4.30
Plums, Green Gage, canned	3.22–3.32
Plums, red	3.60–4.30
Plums, spiced	3.64
Plums, yellow	3.90–4.45
Pollack, filet, broiled	6.72–6.82
Pomegranate	2.93–3.20
Porgy, broiled	6.40–6.49
Pork and beans	5.70
Potatoes	5.40–5.90
Mashed	5.10
Sweet	5.30–5.60
Tubers	5.70
Potato soup	5.90
Prunes, dried, stewed	3.63–3.92
Prune juice	3.95–3.97
Prune, pureed	3.60–4.30
Prune, strained	3.58–3.83
Puffed rice	6.27–6.40
Puffed wheat	5.26–5.77
Pumpkin	4.90–5.50
Quince, fresh, stewed	3.12–3.40
Quince jelly	3.70
Radishes, red	5.85–6.05
Radishes, white	5.52–5.69
Raisins, seedless	3.80–4.10
Rambutan (Thailand)	4.90
Raspberries	3.22–3.95
Raspberries, frozen	3.18–3.26
Raspberries, New Jersey	3.50–3.82
Raspberry jam	2.87–3.17
Razor clams	6.20
Razor shell (sea asparagus)	6.00
Rattan, Thailand	5.20

Table 4   Continued

Item	Approximate pH
Red ginseng	5.50
Red pepper relish	3.10–3.62
Rhubarb, California, stewed	3.20–3.34
Rhubarb	3.10–3.40
Canned	3.40
Rice (all cooked)	
Brown	6.20–6.80
Krispies	5.40–5.73
White	6.00–6.70
Wild	6.00–6.50
Rolls, white	5.46–5.52
Romaine	5.78–6.06
Salmon, fresh, boiled	5.85–6.50
Salmon, fresh, broiled	5.36–6.40
Salmon, Red Alaska, canned	6.07–6.16
Salsa	
Sardines	5.70–6.60
Sardine, Portuguese, in olive oil	5.42–5.93
Satay sauce	5.00
Sauce, enchilada	5.50
Sauce, fish	4.93–5.02
Sauce, Shrimp	7.01–7.27
Sauerkraut	3.30–3.60
Scallion	6.20
Scallop	6.00
Scotch broth	5.92
Sea snail (top shell)	6.00
Shad roe, sautéed	5.70–5.90
Shallots, cooked	5.30–5.70
Sherbet, raspberry	3.69
Sherry-wine	3.37
Shredded Ralston	5.32–5.60
Shredded Wheat	6.05–6.49
Shrimp	6.50–7.00
Shrimp paste	5.00–6.77

(continues)

Table 4 Continued

Item	Approximate pH
Smelts, sautéed	6.67–6.90
Soda crackers	5.65–7.32
Soup	
Broccoli cheese soup, condensed	5.60
Chicken broth, RTS	5.80
Corn soup, condensed	6.80
Cream of celery soup, condensed	6.20
Cream of mushroom, condensed	6.00–6.20
Cream-style corn, condensed	5.70–5.80
Cream of potato soup, condensed	5.80
Cream of shrimp soup, condensed	5.80
Minnestrone soup, condensed	5.40
New England clam chowder, condensed	6.00
Oyster stew, condensed	6.30
Tomato rice soup, condensed	5.50
Soy infant formula	6.60–7.00
Soy sauce	4.40–5.40
Soy bean curd (tofu)	7.20
Soybean milk	7.00
Spaghetti, cooked	5.97–6.40
Spinach	5.50–6.80
Spinach, chopped	5.38–5.52
Spinach, cooked	6.60–7.18
Spinach, frozen, cooked	6.30–6.52
Spinach, pureed	5.50–6.22
Spinach, strained	5.63–5.79
Squash, acorn, cooked	5.18–6.49
Squash, kubbard, cooked	6.00–6.20
Squash, white, cooked	5.52–5.80
Squash, yellow, cooked	5.79–6.00
Squid	6.00–6.50
Sturgeon	6.20
Strawberries	3.00–3.90
Strawberries, California	3.32–3.50
Strawberries, frozen	3.21–3.32
Strawberry jam	3.00–3.40
Straw mushroom	4.90
Sweet potatoes	5.30–5.60
Swiss chard, cooked	6.17–6.78
Tamarind	3.00
Tangerine	3.32–4.48

Table 4 Continued

Item	Approximate pH
Taro syrup	4.50
Tea	7.20
Three-bean salad	5.40
Tofu (soybean curd)	7.20
Tomatillo (resembling cherry tomatoes)	3.83
Tomatoes	4.30–4.90
Tomatoes, canned	3.50–4.70
Tomatoes, juice	4.10–4.60
Tomatoes, paste	3.50–4.70
Tomatoes, puree	4.30–4.47
Tomatoes, strained	4.32–4.58
Tomatoes, wine-ripened	4.42–4.65
Tomato soup, cream of, canned	4.62
Trout, Sea, sautéed	6.20–6.33
Truffle	5.30–6.50
Tuna fish, canned	5.90–6.20
Turnips	5.29–5.90
Turnip, greens, cooked	5.40–6.20
Turnip, white, cooked	5.76–5.85
Turnip, yellow, cooked	5.57–5.82
Vegetable juice	3.90–4.30
Vegetable soup, canned	5.16
Vegetable soup, chopped	4.98–5.02
Vegetable soup, strained	4.99–5.00
Vermicelli, cooked	5.80–6.50
Vinegar	2.40–3.40
Vinegar, cider	3.10
Walnuts, English	5.42
Wax gourd drink	7.20
Water chestnut	6.00–6.20
Watercress	5.88–6.18
Watermelon	5.18–5.60
Wheaties	5.00–5.12
Worcestershire sauce	3.63–4.00
Yams, cooked	5.50–6.81
Yeast	5.65

(continues)

Table 4   Continued

Item	Approximate pH
Yangsberries, frozen	3.00–3.70
Zucchini, cooked	5.69–6.10
Zwieback	4.84–4.94

★ ADDITIONAL INFORMATION ON BSE, vCJD, AND AVIAN INFLUENZA

In this section, we have provided additional information on bovine spongiform encephalopathy (BSE), variant Creutzfeldt-Jakob disease (vCJD), and the avian influenza (bird flu), and avian influenza A (H5N1) virus as communicated by the CDC.

■ ABOUT BSE

Bovine spongiform encephalopathy, commonly known as mad cow disease, is a progressive neurological disorder of cattle that results from infection by an unconventional transmissible agent termed a prion. The nature of the transmissible agent is not well understood. Currently, the most accepted theory is that the agent is a modified form of a normal cell surface component known as prion protein. The pathogenic form of the protein is both less soluble and more resistant to enzyme degradation than the normal form.

Research indicates that the first probable infections of BSE in cows occurred during the 1970s with two cases of BSE being identified in 1986. BSE possibly originated as a result of the feeding of scrapie-containing sheep meat-and-bone meal to cattle. There is strong evidence and general agreement that the outbreak was amplified and spread throughout the Great Britain cattle industry by feeding rendered bovine meat-and-bone meal to young calves.

The BSE epidemic in Great Britain peaked in January 1993 at almost 1,000 new cases per week. By the end of April 2005, more than 184,000 cases of BSE had been confirmed in the United Kingdom alone in more than 35,000 herds.

There has since emerged strong epidemiologic and laboratory evidence for a causal association between variant CJD in humans and the BSE outbreak in cattle. The interval between the most likely period for the initial extended exposure of the population to potentially BSE-contaminated food (1984 to 1986) and the onset of initial variant CJD cases (1994 to 1996) is consistent with known incubation periods for the human forms of the disease.

BSE IN THE UNITED STATES

On June 24, 2005, the U.S. Department of Agriculture announced receipt of final results from The Veterinary Laboratories Agency in Weybridge, England, confirming BSE in a cow that had had conflicting test results in 2004. No parts of the animal entered the human and animal food supply. As a result of this BSE-positive animal, the USDA plans to develop a new laboratory testing protocol to evaluate future inconclusive BSE screening test results.

The only other known case of BSE in the United States was identified in December 2003. On December 23, 2003, the USDA announced a presumptive diagnosis of BSE in an adult Holstein cow from Washington state. This diagnosis was confirmed by an international reference laboratory in Weybridge, England, on December 25. Preliminary trace-back based on an ear-tag identification number suggested that the BSE-infected cow was imported into the United States from Canada in August 2001. This information was later confirmed by genetic testing.

On January 2 and 11, 2005, the Canadian Food Inspection Agency (CFIA) announced the confirmation of BSE in two cows from the province of Alberta. One of the cows was born in October 1996 and the second cow was born in March 1998, after the Canadian government instituted a ruminant feed ban in 1997. No part of these animals has entered the human food supply, according to CFIA.

These two BSE-positive cows bring the total number of BSE-infected cows identified in or linked to Canada to four, including a BSE-positive cow identified in Washington state that was later determined to have originated from Alberta. CDC is in communication with the USDA and will continue to monitor these developments closely. More updated information on the BSE situation in Canada is available from the CFIA BSE Web site, www.inspection.gc.ca/english/anima/heasan/disemala/bseesb/bseesbindexe.shtml.

BSE CONTROL MEASURES

Public health control measures, such as surveillance, culling sick animals, or banning specified risk materials, have been instituted in many countries, particularly in those with indigenous cases of confirmed BSE, in order to prevent potentially BSE-infected tissues from entering the human food supply.

The most stringent control measures include a British program that excludes all animals more than 30 months of age from the human food and animal feed supplies. The program appears to be highly effective.

In June 2000, the European Union Commission on Food Safety and Animal Welfare strengthened the European Union's BSE control measures by requiring all member states to remove specified risk materials from animal feed and human food chains as of October 1, 2000. Other control measures include banning the use of mechanically recovered meat from the vertebral column of cattle, sheep, and goats for human food and BSE testing of all cattle more than 30 months of age destined for human consumption.

■ PREVENTION MEASURES AGAINST BSE SPREAD

To prevent BSE from entering the United States, severe restrictions were placed on the importation of live ruminants, such as cattle, sheep, and goats, and certain ruminant products from countries where BSE was known to exist. These restrictions were later extended to include importation of ruminants and certain ruminant products from all European countries.

Because the use of ruminant tissue in ruminant feed was probably a necessary factor responsible for the BSE outbreak in Great Britain and because of the current evidence for possible transmission of BSE to humans, the USDA instituted a ruminant feed ban in June 1997 that became fully effective as of October 1997.

In late 2001, the Harvard Center for Risk Assessment study of various scenarios involving BSE in the United States concluded that the FDA ruminant feed rule provides a major defense against this disease.

■ BSE/TSE ACTION PLAN OF THE DEPARTMENT OF HEALTH AND HUMAN SERVICES (DHHS)

On August 23, 2001, the Department of Health and Human Services (HHS) issued a department-wide action plan outlining steps to improve scientific understanding of BSE and other transmissible spongiform encephalopathies (TSEs). The action plan has four major components:

★ **Surveillance for human disease** is primarily the responsibility of CDC.

★ **Protection** is primarily the responsibility of the Food and Drug Administration (FDA).

★ **Research** is primarily the responsibility of the National Institutes of Health (NIH).

★ **Oversight** is primarily the responsibility of the Office of the Secretary of the Department of Health and Human Services.

■ FACT SHEET: VARIANT CREUTZFELDT-JAKOB DISEASE (vCJD)

Background

Variant CJD (vCJD) is a rare, degenerative, fatal brain disorder in humans. Although experience with this new disease is limited, evidence to date indicates that there has never been a case of vCJD transmitted through direct contact of one person with another. However, a case of probable transmission of vCJD through transfusion of blood components from an asymptomatic donor who subsequently developed the disease has been reported.

As of June 2005, a total of 177 cases of vCJD had been reported in the world: 156 from Great Britain, 12 from France, 2 from Ireland, and 1 each from Canada, Italy, Japan, the Netherlands, Portugal, Saudi Arabia, and the United States. (Note: The Canadian, one of the Irish, and the Japanese and U.S. cases were reported in persons who visited or resided in the United Kingdom during a key exposure period of the U.K. population to the BSE agent.)

There has never been a case of vCJD that did not have a history of exposure within a country where this cattle disease, BSE, was occurring.

It is believed that the persons who have developed vCJD became infected through their consumption of cattle products contaminated with the agent of BSE. There is no known treatment of vCJD, and it is invariably fatal.

◼ vCJD DIFFERS FROM CLASSIC CJD

This variant form of CJD should not be confused with the classic form of CJD that is endemic throughout the world, including the United States. There are several important differences between these two forms of the disease. The median age at death of patients with classic CJD in the United States, for example, is 68 years, and very few cases occur in persons under 30 years of age. In contrast, the median age at death of patients with vCJD in the United Kingdom is 28 years.

vCJD can be confirmed only through examination of brain tissue obtained by biopsy or at autopsy, but a "probable case" of vCJD can be diagnosed on the basis of clinical criteria developed in the United Kingdom.

The incubation period for vCJD is unknown because it is a new disease. However, it is likely that ultimately this incubation period will be measured in terms of many years or decades. In other words, whenever a person develops vCJD from consuming a BSE-contaminated product, he or she likely would have consumed that product many years or a decade or more earlier.

In contrast to classic CJD, vCJD in the United Kingdom predominantly affects younger people; has atypical clinical features, with prominent psychiatric or sensory symptoms at the time of clinical presentation and delayed onset of neurologic abnormalities, including ataxia within weeks or months, dementia, and myoclonus late in the illness; a duration of illness of at least 6 months; and a diffusely abnormal nondiagnostic electroencephalogram.

The BSE epidemic in the United Kingdom reached its peak incidence in January 1993 at almost 1,000 new cases per week. The outbreak may have resulted from the feeding of scrapie-containing sheep meat-and-bone meal to cattle. There is strong evidence and general agreement that the outbreak was amplified by feeding rendered bovine meat-and-bone meal to young calves.

◼ U.S. SURVEILLANCE FOR CJD

The Centers for Disease Control and Prevention (CDC) monitors the trends and current incidence of CJD in the United States by analyzing death certificate information from U.S. multiple cause-of-death data, compiled by the National Center for Health Statistics. From 1979 to 2003, the average annual age-adjusted death rates of CJD (not vCJD) have remained relatively stable. In addition, deaths from non-iatrogenic CJD (not vCJD) in persons aged 30 years or less in the United States remain extremely rare (less than 5 cases per 1 billion per year). In contrast, in Great Britain, over half of the patients who died with vCJD were in this young age group.

In addition, the CDC collects, reviews, and, when indicated, actively investigates reports by health care personnel or institutions of possible CJD or vCJD cases. Also, from 1996 to 1997, the CDC established, in collaboration with the American Association of Neuropathologists, the National Prion Disease Pathology Surveillance

Center at Case Western Reserve University, which performs special diagnostic tests for prion diseases, including postmortem tests for vCJD.

■ ABOUT AVIAN INFLUENZA (BIRD FLU)

Bird flu is an infection caused by avian (bird) influenza (flu) viruses. These flu viruses occur naturally among birds. Wild birds worldwide carry the viruses in their intestines, but usually do not get sick from them. However, bird flu is very contagious among domesticated birds and can make some, including chickens, ducks, and turkeys, very sick and kill them. Bird flu viruses do not usually infect humans, but several cases of human infection with bird flu viruses have occurred since 1997.

There are many different subtypes of type A flu viruses. These subtypes differ because of certain proteins on the surface of the flu A virus (hemagglutinin [HA] and neuraminidase [NA] proteins). There are 16 different HA subtypes and 9 different NA subtypes of flu A viruses. Many different combinations of HA and NA proteins are possible. Each combination results in a different subtype. All subtypes of flu A viruses can be found in birds. However, when we talk about bird flu viruses, we are referring to those flu A subtypes that continue to occur mainly in birds. They do not usually infect humans, even though we know they can do so. When we talk about "human flu viruses," we are referring to those subtypes that occur widely in humans. There are only three known subtypes of human flu viruses (H1N1, H1N2, and H3N2); it is likely that some genetic parts of current human flu A viruses came from birds originally. Flu A viruses are constantly changing, and they might adapt over time to infect and spread among humans.

Symptoms of bird flu in humans have ranged from typical flulike symptoms (fever, cough, sore throat and muscle aches) to eye infections, pneumonia, severe respiratory diseases (such as acute respiratory distress), and other severe and life-threatening complications. The symptoms of bird flu may depend on which virus caused the infection.

Infected birds shed flu virus in their saliva, nasal secretions, and feces. Susceptible birds become infected when they have contact with contaminated excretions or surfaces that are contaminated with excretions. It is believed that most cases of bird flu infection in humans have resulted from contact with infected poultry or contaminated surfaces.

Studies suggest that the prescription medicines approved for human flu viruses would work in preventing bird flu infection in humans. However, flu viruses can become resistant to these drugs, so these medications may not always work.

The risk from bird flu is generally low to most people because the viruses occur mainly among birds and do not usually infect humans. However, during an outbreak of bird flu among poultry (domesticated chicken, ducks, turkeys), there is a possible risk to people who have contact with infected birds or surfaces that have been contaminated by infected birds. The current outbreak of avian influenza A (H5N1) among poultry in Asia (see next section) is an example of a bird flu outbreak that has caused human infections and deaths. In such situations, people should avoid contact with infected birds or contaminated surfaces, and should be careful when handling and cooking poultry. For more information about avian influenza and food safety issues, visit the World Health Organization Web site at www.who.int/foodsafety/micro/avian/en.

■ ABOUT AVIAN INFLUENZA A (H5N1) VIRUS

Influenza A virus—also called H5N1 virus—is an influenza A virus subtype that occurs mainly in birds. Like all bird flu viruses, H5N1 virus circulates among birds worldwide, is very contagious among domesticated birds, and can be deadly.

Outbreaks of influenza H5N1 have occurred among poultry in eight countries in Asia (Cambodia, China, Indonesia, Japan, Laos, South Korea, Thailand, and Vietnam) during late 2003 and early 2004. As of February 2006, there are 20 countries affected by the avian flu. At that time, more than 100 million birds in the affected countries either died from the disease or were killed in order to try to control the outbreak. By March 2004, the outbreak was reported to be under control. Beginning in late June 2004, however, new deadly outbreaks of influenza H5N1 among poultry were reported by several countries in Asia (Cambodia, China, Indonesia, Malaysia [first-time reports], Thailand, and Vietnam). It is believed that these outbreaks are ongoing. Human infections of influenza A (H5N1) have been reported in Thailand, Vietnam, and Cambodia.

While the H5N1 virus does not usually infect humans, in 1997, the first case of bird-to-human spread was seen during an outbreak of bird flu in poultry in Hong Kong. The virus caused severe respiratory illness in 18 people, 6 of whom died. Since that time, there have been other cases of H5N1 infection among humans. Most recently, human cases of H5N1 infection have occurred in Thailand, Vietnam, and Cambodia. The death rate for these reported cases has been about 50 percent. Most of these cases occurred from contact with infected poultry or contaminated surfaces; however, it is thought that a few cases of human-to-human spread of H5N1 have occurred.

So far, spread of H5N1 virus from person to person has been rare and spread has not continued beyond one person. However, because all influenza viruses have the ability to change, scientists are concerned that the H5N1 virus could one day be able to infect humans and spread easily from one person to another. Because these viruses do not commonly infect humans, there is little or no immune protection against them in the human population. If the H5N1 virus were able to infect people and spread easily from person to person, an "influenza pandemic" (worldwide outbreak of disease, see www.cdc.gov/flu/avian/gen-info/pandemics.htm) could begin. No one can predict when a pandemic might occur. However, experts from around the world are watching the H5N1 situation in Asia very closely and are preparing for the possibility that the virus may begin to spread more easily and widely from person to person.

The H5N1 virus in Asia that has caused human illness and death is resistant to amantadine and rimantadine, two antiviral medications commonly used for influenza prevention. Two other antiviral medications, oseltamivir, and zanamivir, would probably work to treat flu caused by the H5N1 virus, though studies still need to be done to prove that they work.

There currently is no vaccine to protect humans against the H5N1 virus that is being seen in Asia. However, vaccine development efforts are under way. Research studies to test a vaccine to protect humans against H5N1 virus began in April 2005. (Researchers are also working on a vaccine against H9N2, another bird flu virus subtype.) For more information about the H5N1 vaccine development process, visit the National Institutes of Health Web site at www2.niaid.nih.gov/Newsroom/Releases/flucontracts.htm.

The current risk to Americans from the H5N1 bird flu outbreak in Asia is low. The strain of H5N1 virus found in Asia has not been found in the United States. There have been no human cases of H5N1 flu in the United States. It is possible that travelers returning from affected countries in Asia could be infected. Since February 2004, medical and public health personnel have been watching closely to find any such cases.

In February 2004, CDC provided U.S. health departments with recommendations for enhanced surveillance ("detection") in the U.S. of avian influenza A (H5N1). Follow-up messages (Health Alert Network) were sent to the health departments in August 2004 and February 2005, both reminding health departments about how to detect (domestic surveillance), diagnose, and prevent the spread of avian influenza A (H5N1). It also recommended measures for laboratory testing for H5N1 virus. CDC currently advises that travelers to countries in Asia with known outbreaks of influenza A (H5N1) avoid poultry farms, contact with animals in live food markets, and any surfaces that appear to be contaminated with feces from poultry or other animals.

■ POSSIBLE H5N1 FLU PANDEMIC

The CDC is taking part in a number of pandemic prevention and preparedness activities, including:

★ Working with the Association of Public Health Laboratories on training workshops for state laboratories on the use of special laboratory (molecular) techniques to identify H5 viruses

★ Working with the Council of State and Territorial Epidemiologists and others to help states with their pandemic planning efforts

★ Working with other agencies such as the Department of Defense and the Veterans Administration on antiviral stockpile issues

★ Working with the World Health Organization (WHO) and Vietnamese Ministry of Health to investigate influenza H5N1 in Vietnam and to provide help in laboratory diagnostics and training to local authorities

★ Performing laboratory testing of H5N1 viruses

★ Starting a $5.5 million initiative to improve influenza surveillance in Asia

★ Holding or taking part in training sessions to improve local capacities to conduct surveillance for possible human cases of H5N1 and to detect influenza A H5 viruses by using laboratory techniques

★ Developing and distributing reagents kits to detect the currently circulating influenza A H5N1 viruses

★ Working together with WHO and the National Institutes of Health (NIH) on safety testing of vaccine seed candidates and to develop additional vaccine virus seed candidates for influenza A (H5N1) and other subtypes of influenza A virus

For more information, visit www.cdc.gov/flu, or call CDC at 800-CDC-INFO (English and Spanish) or 888-232-6348 (TTY)

ACCEPTABLE LEVEL   The presence of a food safety hazard—biological, chemical, or physical—at which levels are low enough not to cause an illness or injury.

ACID   A substance with a pH of less than 7.0.

ACTIVE MANAGERIAL CONTROL   The purposeful incorporation of specific actions or procedures by management in the operation of their business to attain control over the five foodborne illness risk factors identified by the CDC.

ADULTERATED   Food that contains a poisonous or deleterious substance that causes it to be hazardous or unfit for human consumption.

AEROBIC   Able to reproduce and live only in the presence of free oxygen.

ALKALI   A substance with a pH of more than 7.0.

ANAEROBIC   Able to reproduce and live in the absence of free oxygen.

APPROVED SOURCE   The regulatory authority deemed an acceptable supplier based on a determination of conformity with laws, statutes, regulations, principles, practices, and generally recognized standards of operation that protect public health and safety.

BACTERIA   Single-cell microorganisms, usually classified as the simplest of plants.

BIOLOGICAL HAZARD   Danger to food from disease-causing microorganisms known as pathogens, poisonous plants, mushrooms, and fish that carry harmful toxins.

BIO-TERRORISM   Intentionally infecting people to cause illness and death by the spread of highly contagious diseases like smallpox, anthrax, botulism, plague, and viral hemorrhagic fevers.

BOILING-POINT   The temperature at which a liquid changes to a gas. The boiling point of water is 212°F (100°C) at sea level.

CCP   See Critical Control Point.

CDC   Centers for Disease Control and Prevention.

CFR   See Code of Federal Regulations.

CHEMICAL HAZARD   Danger to food posed by chemical substances, especially toxic metals, pesticides, and food additives.

CIP   Cleaned in place. This is the process of washing, rinsing, and sanitizing equipment through a piping/mechanical system used in frozen dessert machines.

CLEAN   Free of visible soil.

CODE OF FEDERAL REGULATIONS (CFR)   The application of codes to the general and permanent rules established by federal agencies and departments and published in the Federal Register.

CONTAMINATION   The presence in food of potentially harmful substances, including microorganisms (bacteria, virus, parasites), chemicals (pesticides, toxic metals), and physical objects (hair, dirt, glass).

CONTROL MEASURE   Any action or activity that can be used to prevent, eliminate, or reduce an identified biological, chemical, or physical hazard. Control measures determined to be essential for food safety are applied at critical control points in the flow of food.

CONTROL POINT (CP)   Any step in the flow of food when control can be applied to prevent, reduce, or eliminate a biological, chemical, or physical hazard. Loss of control at this point will not result in unsafe or high-risk level of food.

CONTROLLED ATMOSPHERE PACKAGING (CAP) FOODS   Using CAP, a package of food is modified so that until the package is opened, its composition is different from air, and continuous control of that atmosphere is maintained, such as by using oxygen scavengers (chemicals placed directly into the packaging wall that absorbs oxygen that permeates into the package over time) or a combination of total replacement of oxygen, nonrespiring foods (i.e., meat and seafood), and impermeable packaging material. The food product is packaged in a laminate or film, following which the atmosphere inside the pack is controlled.

COOK CHILL PROCESSING (CC)   Cook chill packaging, in which cooked food is hot-filled into impermeable bags that have the air expelled and are then sealed or crimped closed. The bagged food is rapidly chilled and refrigerated at temperatures that inhibit the growth of pathogens.

CORRECTIVE ACTION   An activity that is taken by a person whenever a critical limit is not met or a deviation occurs.

CP   See Control Point.

CRITERION   A requirement on which a judgment or decision can be based.

CRITICAL CONTROL POINT (CCP)   An operational step, point, or procedure in a food preparation process at which control can be applied and is essential to prevent or eliminate a biological, chemical, and physical hazard or reduce it to acceptable levels. A loss of control at this point results in unsafe and high-risk levels in food.

CRITICAL LIMIT (CL)   A criterion of one or more prescribed parameters that must be met to ensure that a CCP effectively controls a hazard.

CROSS-CONTAMINATION   The transfer of harmful substances or disease-causing microorganisms from one surface to another by hands, food-contact surfaces, sponges, cloth towels, equipment, storage, and utensils.

DANGER ZONE   The temperature range between 41°F (5°C) and 135°F (57°C) that favors the growth of pathogenic microorganisms.

DATE MARKING   The practice of indicating the date or day by which all food should be consumed, sold, or discarded.

DECLINE / DEATH PHASE   The phase of bacteria growth, following the stationary phase, in which the rate of death within the colony exceeds the rate of reproduction and the number of living cells begin to decrease.

DELETERIOUS SUBSTANCE   A substance that is harmful or injurious.

DEVIATION   The failure to meet a required critical limit for a critical control point.

ESCHERICHIA COLI (E. coli)   A bacteria associated with cattle that causes human illness when undercooked, contaminated ground beef is eaten. Infection can also occur after drinking raw milk or by contact with sewage-contaminated water. Food service employees with E. coli must report this infection to the proper health authorities.

EXCLUDE To prevent a person from working as a food employee or entering a food establishment except for those areas open to the general public.

FATTOM An acronym for food, acidity, time, temperature, oxygen, and moisture. The letters in FATTOM stand for the conditions needed for microorganisms to grow.

FDA United States Food and Drug Administration

FISH A fresh or saltwater finfish, crustaceans, and other forms of aquatic life (including alligator, frog, aquatic turtle, jellyfish, sea cucumber, and sea urchin) and all mollusks, if intended for human consumption.

FOOD Raw, cooked, or processed edible substance, ice, beverage, chewing gum, or ingredient used or intended for use or for sale in whole or in part for human consumption.

FOOD ALLERGY Condition caused by a reaction to naturally occurring protein in a food or a food ingredient. **Major food allergens** and the **Big 8** are the foods that account for 90 percent or more of all food allergies. They are shellfish (crab, lobster, or shrimp); fish (bass, flounder, or cod); peanuts; tree nuts (almonds, pecans, chestnuts, pistachios, Brazil nuts, etc.); milk; eggs; soy/tofu; and wheat.

FOOD DEFENSE The protection of food products from intentional adulteration/contamination.

FOOD ESTABLISHMENT An operation at the retail or foodservice level that serves or offers food directly to the consumer and that, in some cases, includes a production, storage, or distributing operation that supplies the direct-to-consumer operation.

FOOD POISONING A general term for intoxication or infection caused by consumption of contaminated food.

FOOD PREPARATION PROCESS A series of operational steps conducted to produce a food ready to be consumed.

FOOD SECURITY Best defined by the **World Health Organization** (WHO), food security is "the implication that all people at all times have both physical and economic access to enough food for an active, healthy life." Internationally, food security is defined as a 2-year supply of food for a particular country.

FOODBORNE ILLNESS A sickness resulting from the consumption of foods or beverages contaminated with disease-causing microorganisms (pathogens), chemicals (pesticides), or other harmful substances (glass).

FOODBORNE OUTBREAK The occurrence of two or more cases of illness resulting from the ingestion of a common food.

GAME ANIMAL In general, an animal such as bison, deer, elk, rabbit, raccoon, and squirrel. Game animals are not ratites or livestock.

HACCP Hazard Analysis and Critical Control Point.

HACCP PLAN A written document that is based on the principles of HACCP and describes the procedures to be followed to ensure the control of a specific process or procedure.

HACCP SYSTEM The result of implementing the HACCP principles in an operation that has foundational comprehensive, prerequisite programs in place. A HACCP system includes all prerequisite programs and the HACCP plan, including all seven HACCP principles.

HACCP TEAM A group of people who are responsible for developing and implementing a HACCP plan.

HAZARD A biological, physical, or chemical property that may cause a food to be unsafe for human consumption.

HAZARD ANALYSIS AND **CRITICAL CONTROL POINT (HACCP)** A prevention-based food safety system that identifies and monitors specific food safety hazards that can adversely affect the safety of food products.

HEPATITIS A   A virus that can be transmitted through direct contact with a hepatitis A infected person, or ingestion of hepatitis A contaminated food or water. Often seen with a jaundice condition, a foodservice employee diagnosed with this virus must report it to the proper health authorities.

HERMETICALLY SEALED CONTAINER   A container that is designed to keep microorganisms out in such products as low-acid canned foods.

HIGHLY SUSCEPTIBLE POPULATION (HSP)   Persons who are more likely than other populations to experience foodborne disease because they are either immunocompromised, preschool-age children (infants or toddlers), or older adults.

HOAX   A false claim of damage due to food contamination or an intentional contamination of food and then claiming damage.

HYGIENE   Practices necessary for establishing and maintaining good health.

ICE POINT   The temperature at which a liquid changes to a solid. The ice point of water is 32°F (0°C).

INCUBATION PERIOD   The phase in the course of an infection between the invasion of the host by the pathogen and the appearance of the symptoms of illness.

INFECTION   Disease caused by invasion of living pathogenic organisms, which multiply within the body, causing illness.

INTOXICATION   Disease caused by consumption of poisons (toxins), which may be chemical, naturally occurring in food, or produced by pathogenic microorganisms.

JAUNDICE   A condition that causes the skin and eyes to yellow.

LAG PHASE   The period of bacterial growth following transfer to a new environment, when adaptation to new conditions takes place and there is little or no increase in the number of cells in the colony.

LOG PHASE   The period of bacterial growth following the lag phase, when multiplication rate is constant and rapid and the number of cells in the colony increases exponentially.

MEAT   The flesh of animals used as food including the dressed flesh of cattle, swine, sheep, or goats and other edible animals, except fish, poultry, and wild-game animals.

MICROBE   A general term for microscopic organisms, particularly pathogens.

MICROORGANISM   A form of life that can be seen only with a microscope, including bacteria, viruses, yeast, and single-celled animals.

MODIFIED ATMOSPHERE PACKAGED (MAP) FOODS   Food that is partially processed or lightly cooked before being put into a container and sealed. The MAP process uses special gases that control reduction in the proportion of oxygen, total replacement of oxygen, or an increase in the proportion of other gases such as carbon dioxide or nitrogen.

MOLLUSCAN SHELLFISH   Any edible species of raw fresh or frozen oysters, clams, mussels, and scallops or edible portions thereof, except when the scallop product consists only of the shucked adductor muscle.

MONITORING   The act of observing and making measurements to help determine if critical limits are being met and maintained.

NATIONAL SHELLFISH SANITATION PROGRAM (NSSP)   The voluntary system by which regulatory authorities for shellfish-harvesting waters and shellfish processing and transportation and the shellfish industry implement specified controls to ensure that raw and frozen shellfish are safe for human consumption.

NOROVIRUS   A gastrointestinal virus that is commonly called the "Norwalk-like virus," "small round-structured virus," and "winter vomiting disease," resulting in nausea, diarrhea, vomiting, and stomach cramps. Because it is highly contagious, a norovirus must be reported to the proper health authorities.

NSSP   National Shellfish Sanitation Program.

OPERATIONAL STEP   An activity or stage in the flow of food through a food establishment, such as purchasing, receiving, storage, preparation, cooking, holding, cooling, reheating, and serving.

ORGANISM   An individual living thing.

PARASITE   An organism that lives on or in another, usually larger, host organism in a way that harms or is of no advantage to the host.

PATHOGEN   A microorganism (bacteria, parasites, viruses, or fungi) that causes disease in humans.

PATHOGENIC   Disease-causing microorganisms.

PCO   Pest control operator (licensed).

PERSONAL HYGIENE   Individual cleanliness and habits.

PERSON IN CHARGE   The individual present at a food establishment who is responsible for the operation at the time of inspection.

pH   The measure of the acidity of a product. Key: pH 0 to 7 is acidic; pH of 7 is neutral; and pH 7 to 14 is alkaline.

POTENTIALLY HAZARDOUS FOOD/TIME/TEMPERATURE CONTROL FOR SAFETY OF FOOD (PHF/TCS)   A food that requires TTC to limit pathogenic microorganism growth or toxin formation.

POULTRY   Any domesticated bird (chicken, turkey, duck, geese, guineas, ratites, squab) or game bird (pheasant, partridge, quail, growe, pigeon).

PREREQUISITE PROGRAMS   Procedures, including standard operating procedures (SOPs), that address basic operational and sanitation conditions in an establishment.

PREVENTIVE MEASURE   Physical, chemical, or other factors that can be used to control an identified health hazard.

PROCEDURAL STEP   An individual activity in applying the contents of this book to a food establishment's operations.

PROCESS APPROACH   A method of categorizing food operations into one of three categories:

Simple/no-cook step: Food preparation with no-cook step wherein ready-to-eat food is received, stored, prepared, held, and served.

Same-day service: Food preparation for same-day service wherein food is received, stored, prepared, cooked, held, and served.

Complex food preparation: Complex food preparation wherein food is received, stored, prepared, cooked, cooled, reheated, hot-held, and served.

RATITE   A flightless bird like an ostrich, emu, and rhea.

READY-TO-EAT (RTE) FOOD   RTE foods include the following:

- ★ Raw animal foods that have been properly cooked
- ★ Fish intended for raw consumption that has been frozen to destroy parasites
- ★ Raw fruits and vegetables that are washed
- ★ Fruits and vegetables that are cooked for hot holding
- ★ Plant food for which further washing, cooking, or other processing is not required for food safety, and from which rinds, peels, husks, or shells, if naturally present, are removed
- ★ Substances derived from plants such as spices, seasonings, and sugar; a bakery item such as bread, cakes, pies, fillings, or icing for which further cooking is not required for food safety

* Dry, fermented sausages, such as dry salami or pepperoni
* Salt-cured meat and poultry products, such as prosciutto ham, country-cured ham, and Parma ham
* Dried meat and poultry products, such as jerky or beef sticks, and low-acid foods that have been thermally processed and packaged in hermetically sealed containers.

RECORD   A documentation of monitoring observations and verification activities.

REDUCED OXYGEN PACKAGING (ROP)   Encompasses a large variety of packaging methods where the internal environment of the package contains a controlled oxygen level (typically 21 percent at sea level), including vacuum packaging (VP), modified atmosphere packaging (MAP), controlled atmosphere packaging (CAP), cook chill processing (CC), and sous vide (SV). Using ROP methods in food establishments has the advantage of providing extended shelf life to many foods because it inhibits spoilage organisms that are typically aerobic.

REGULATORY AUTHORITY   A federal, state, local, or tribal enforcement body or authorized representative having jurisdiction over the food establishment.

RESTRICT   To limit the activities of a food employee so that there is no risk of transmitting a disease that is transmissible through food and the food employee does not work with exposed food, clean equipment, utensils, linens, and unwrapped single-service or single-use articles.

RISK ANALYSIS   An estimate of the likely occurrence of a hazard.

RISK CONTROL PLAN (RCP)   A concisely written management plan developed by the retail or foodservice operator with input from the health inspector that describes a management system for controlling specific out-of-control risk factors.

RISK FACTOR   One of the broad categories of contributing factors to foodborne illness outbreaks, as identified by the CDC, that directly relates to foodborne safety concerns within retail and foodservice establishments. The five factors are poor personal hygiene, inadequate cooking temperatures, improper holding temperatures, contaminated equipment, and food from unsafe sources.

SALMONELLA   A bacteria that can cause diarrhea, fever, and stomach pain in people who have eaten food or contact animals with the salmonella bacteria. This could be highly contagious, so foodservice employees with this disease should report it to the proper health authorities.

SANITARY   Free of disease-causing organisms and other harmful substances.

SANITIZATION   The reduction of the number of pathogenic microorganisms on a surface to levels accepted as safe by regulatory authorities.

SEVERITY   The seriousness of the effect(s) of a hazard.

SHIGELLOSIS   A bacterial infection causing severe diarrhea that can pass from person to person or from eating contaminated food. Food may become contaminated by infected food handlers who do not properly wash hands after using the restroom. Flies and sewage-contaminated water are other sources. Any food handler with shigellosis must report it to the proper health authorities.

SOP   Standard operating procedure.

SOUS VIDE (SV)   Raw or partially cooked food that is packaged in a hermetically sealed, impermeable bag; cooked in the bag; rapidly chilled; and refrigerated at temperatures that inhibit the growth of pathogens.

SPORE   A very tough, dormant form of certain bacterial cells that is very resistant to desiccation, heat, and a variety of chemical and radiation treatments that are otherwise lethal to vegetative cells.

SPORE FORMER   A bacterium capable of producing spores under adverse conditions. Spore formers in food include *Clostridium botulinum*, *Bacillus cereus,* and *Clostridium perfringens*.

STANDARD OPERATING PROCEDURE (SOP)   A written method of controlling a practice in accordance with predetermined specifications to obtain a desired outcome.

STATIONARY PHASE   The period of bacterial growth, following the log phase, in which the number of bacterial cells remains more or less constant, as cells compete for space and nourishment.

TC   Temperature control.

TEMPERATURE MEASURING DEVICE   A thermometer, thermocouple, thermistor, or other device for measuring the temperature of food, air, or water.

TOXIGENIC MICROORGANISMS   Pathogenic bacteria that cause foodborne illness in humans due to the ingestion of poisonous toxins produced in food.

TOXIN   A poisonous substance that may be found in food.

USDA   U. S. Department of Agriculture.

VACUUM PACKAGING (VP)   The process in which air is removed from a package of food and the package is hermetically sealed so that a vacuum remains inside the package.

VALIDATION   That element of verification focused on collecting and evaluating scientific and technical information to determine if the HACCP plan, when properly implemented, will effectively control the hazards.

VARIANCE   A written waiver issued and authorized by a regulatory agency.

VEGETATIVE CELL   A bacterial cell that is capable of actively growing.

VERIFICATION   Ensuring that monitoring and other functions of a HACCP plan are being properly implemented.

VIRUS   The smallest of microorganisms that is dependent on a living host cell to survive and multiply, and therefore cannot multiply in or on food.

WATER ACTIVITY (a_w)   The amount of water available in the product to allow bacteria to live and grow. Scientifically, it is the quotient of the water vapor pressure of the substance, divided by the vapor pressure of pure water at the same temperature.

ZOONOTIC DISEASE   A disease that is communicable from animals to humans such as BSE (mad cow), avian flu, E. coli, salmonella, rabies, and malaria.

AFDO (Association of Food and Drug Officials)

★ www.afdo.org

AFDO is an international nonprofit organization that has six regional affiliates that serve as resources for you and your foodservice operation:

★ CASA—Central Atlantic States Association of Food and Drug Officials

www.casafdo.org

★ AFDOSS—Association of Food and Drug Officials of the Southern States

www.afdoss.org

★ MCAFDO—Mid-Continental Association of Food and Drug Officials

www.mcafdo.org

★ NCAFDO—North Central Association of Food and Drug Officials

www.ncafdo.org

★ WAFDO—Western Association of Food and Drug Officials

www.wafdo.org

★ NEFDOA—North East Food and Drug Officials Association

www.nefdoa.org

American State Health Officials: 202-371-9090

★ www.statepublichealth.org

Centers for Disease Control and Prevention: 800-CDC-INFO or 888-232-6348

★ www.cdc.gov

Center for Safety and Applied Nutrition: 1-888-SAFEFOOD

★ www.cfsan.fda.gov/~dms/foodcode.html

★ www.cfsan.fda.gov/~dms/secgui11.html

★ Center for Food Safety and Applied Nutrition,

Office of Compliance
Dr. John E. Kvenberg, Deputy Director, OC
U.S. Food and Drug Administration
HFS-600
5100 Paint Branch Parkway
College Park, MD 20740-3835

Department of Homeland Security: 202-324-0001

★ www.ready.gov or www.dhs.gov

FBI / Federal Bureau of Investigations: 202-456-1111

★ www.fbi.gov

FDA: 1-888-INFO-FDA (1-888-463-6332) or 301-443-1240

★ www.fda.gov

★ United States Food and Drug Administration

Center for Food Safety and Applied Nutrition, Retail Food Protection Team
U.S. Food and Drug Administration
FDA, HFS-627
5100 Paint Branch Parkway
College Park, MD 20740-3835

FDA Publications and Federal Regulations

★ **FDA Food Code,** current edition, may be purchased from the U.S. Department of Commerce, National Technical Information Service, via telephone: 703-487-4650 or electronically via the FDA Web site at www.cfsan.fda.gov/~dms/foodcode.html.

★ **Fish and Fishery Products—Code of Federal Regulations,**

Title 21, Part 123 Fish and Fishery Products.

Fish and Fishery Products Hazards and Controls Guide, Third Edition, June 2001. Food and Drug Administration, Washington, D.C. May be purchased from: National Technical Information Service, U.S. Department of Commerce, 703-487-4650

★ The **Fish and Fishery Products Hazards and Controls Guide** is also available electronically at www.cfsan.fda.gov/~comm/haccpsea.html. Single copies may be obtained as long as supplies last from FDA district offices and from:

U.S. Food and Drug Administration
Office of Seafood
5100 Paint Branch Parkway
College Park, MD 20740-3835

★ **National Shellfish Sanitation Program Model Ordinance for Molluscan Shellfish,** available on the FDA/CFSAN Web site at www.cfsan.fda.gov/~ear/nsspo-toc.html or may be purchased from:

National Technical Information Service
U.S. Department of Commerce
703-487-4650

★ **Report of the FDA Retail Food Program Database of Foodborne Illness Risk Factors**, available on the FDA/CFSAN Web site at www.cfsan.fda.gov/~dms/retrsk.html

★ **FDA Report on the Occurrence of Foodborne Illness Risk Factors in Selected Institutional Foodservice, Restaurant, and Retail Food Store Facility Types,** available on the FDA/CFSAN Web site at www.cfsan.fda.gov/~dms/retrsk2.html

FDA Regional Field Offices (Regional Retail Food Specialists)

★ **Northeast** (Maine, New Hampshire, Massachusetts, Vermont, Rhode Island, Connecticut, and New York):

158-15 Liberty Avenue, HFR-NE4
Jamaica, NY 11433-1034
718-662-5621 Fax 718-662-5434

or

One Montvale Avenue, HFR-NE250
Stoneham, MA 02180-3542
781-596-7700 Fax 781-596-7896

★ **Central—Mid-Atlantic** (New Jersey, Delaware, District of Columbia, Maryland, Pennsylvania, Virginia, West Virginia, Kentucky, and Ohio):

101 West Broad Street, Suite 400
Falls Church, VA 22046
703-235-8440 ext. 502

★ **Central—Midwest** (Illinois, Indiana, Michigan, Minnesota, North Dakota, South Dakota, and Wisconsin):

20 North Michigan Ave., Suite 50, HFR-MW15
Chicago, IL 60602-4811
312-353-9400 Fax 312-886-1682

or

240 Hennepin Avenue
Minneapolis, MN 55401
612-334-4100 ext. 115 Fax 612-334-4134

★ **Southeast** (Alabama, Florida, Georgia, Louisiana, Mississippi, North Carolina, South Carolina, Puerto Rico, Tennessee, and Virgin Islands):

60 8th Street, N.E. HFR-SE13
Atlanta, GA 30309-3959
404-253-1200 ext. 1265, 1267, 1268, 1273 Fax 404-253-1207

★ **Southwest** (Arkansas, Oklahoma, Texas, Colorado, New Mexico, Wyoming, Utah, Missouri, Kansas, Iowa, and Nebraska):

4040 N. Central Expressway, Suite 900, HFR-SW16
Dallas, TX 75204
214-253-4948, 4947, 4945 Fax 214-253-4960

or

11510 W. 8th Street, HFR-SW36
Lenexa, KS 66285-5905
913-752-2401 Fax 913-752-2487

or

Building 20, Denver Federal Center
P.O. Box 25087
Denver, CO 80225-0087
303-236-3026 Fax: 303-236-3551

★ **Pacific** (Alaska, Arizona, American Samoa, California, Hawaii, Guam, Nevada, Idaho, Oregon, Washington, and Montana):

Office of Regional Director - Pacific Region
Oakland Federal Bldg., HFR-PA16
1301 Clay Street, Suite 1180N
Oakland, CA 94612-5217
510-637-3960 ext. 27 Fax: 510-637-3976

or

51 West Third Street
Tempe, AZ 85281
480-829-7396 ext. 35 Fax: 480-829-7677

or

9780 SW Nimbus Avenue
Beaverton, OR 97008-7163
503-671-9711 ext. 16 Fax: 503-671-9445

FSIS / Food Safety and Inspection Service: 1-800-333-1284

★ www.fsis.usda.gov/ (FSIS Main Page)

★ www.fsis.usda.gov/oa/topics/securityguide.htm

★ www.fsis.usda.gov/oa/recalls/rec_intr.htm (Recall Information Center)

★ www.fsis.usda.gov/oa/pubs/recallfocus.htm

★ www.cfsan.fda.gov/~lrd/recall2.html

★ **United States Department of Agriculture Food Safety and Inspection Service**

Office of the Director
USDA FSIS PPID/HACCP
Room 6912, Suite 6900E
1099 14th Street, N.W.
Washington, DC 20250-3700
202-501-7319 Fax: 202-501-7639

General Accounting Office: 202-512-4800

★ www.gao.gov/new.items/rc00003.pdf

International Association for Food Protection: 800-369-6337

6200 Aurora Avenue, Suite 200W
Des Moines, IA 50322-2864

★ www.foodprotection.org

Regulatory Services for Approved Suppliers

★ CDC Food Safety Office—404-639-2213 or www.cdc.gov

★ EPA—202-272-0167 or www.epa.gov

★ FSIS—888-674-6854 or www.fsis.usda.gov

★ FDA—888-463-6332 or www.cfsan.fda.gov

★ USDC—seafood.nmfs.noaa.gov (approved list of fish products)

US Department of Agriculture: 202-720-3631 or 1-800-233-3935

★ www.usda.gov/wps/portal/usdahome (home page)

★ schoolmeals.nal.usda.gov/Safety/biosecurity.pdf

★ www.nfsmi.org/Information/e-readinessguide.pdf

★ www.nfsmi.org/Information/recallmanual.pdf